Tables, Data and Formulae for
ENGINEERS

compiled by

A. Greer, C.Eng., M.R.Ae.S.

Senior Lecturer in Engineering,
City of Gloucester College of Technology

and

D. J. Hancox, B. Sc., F.I.M.A.

Head of Department of Mathematics,
Coventry Technical College

Stanley Thornes (Publishers) Ltd

Text © A Greer, D Hancox 1977   Diagrams © Stanley Thornes (Publishers) Ltd., 1977

The authors and publishers gratefully acknowledge permission to reproduce copyright material in the form of the table of logarithms, antilogarithms, natural logarithms, reciprocals and normal distribution, as originally compiled by Messrs J White, A Yeats and G Skipworth for TABLES FOR STATISTICIANS published also by Stanley Thornes (Publishers) Ltd.

First published by Stanley Thornes (Publishers) Ltd.
EDUCA House, 32 Malmesbury Road, Kingsditch Estate
CHELTENHAM   England

ISBN 0 85950 023 3

British Standard Specification numbers referred to are those ruling at the time of publication.

References to British Standards are reproduced by kind permission of the British Standards Institution from whom can be obtained the full standard.

The EN numbers shown within the Properties of Materials section were those formerly used, and are for information only. They should not be used for specification purposes.

Typeset by Avontype, Bristol
Printed by The Pitman Press, Bath

Contents

Mathematical signs and abbreviations

Symbol	Term	Symbol	Term
[{( )}]	brackets	$\lim y$	limit of y
$+$	plus	$\to a$	approaches a
$-$	minus	∞	infinity
$\pm$	plus or minus	Σ	sum of
$\|a-b\|$	modulus of difference between a and b	Π	product of
$\times$ or $\cdot$	multiplied by	$\sqrt{x}, x^{\frac{1}{2}}$	square root of x
$\div$ or $/$	divided by	$x^{\frac{1}{3}}$	cube root of x
$=$	is equal to	e	base of natural logarithms
$\neq$	is not equal to	$\log_a x$	logarithm to the base a
$\equiv$	is identical with	$\ln x, \log_e x$	natural logarithm of x
$\hat{=}$	corresponds to	$\lg x, \log_{10} x$	common logarithm of x
$\approx$	is approximately equal to	antilog	antilogarithm
$\sim$	is asymptotically equal to	$\exp x, e^x$	exponential function of x
$\propto$	varies directly as	$n!$	factorial n
$>$	is greater than	$\binom{n}{p}, {}^nC_p$	binomial coefficient
$<$	is less than		
$\geqslant$	is equal to or greater than	Δ, δ	increment or finite difference operator
$\leqslant$	is equal to or less than		
$\gg$	is much greater than	D	operator $\dfrac{d}{dx}$
$\ll$	is much less than		
i, j	complex number $i = j = \sqrt{-1}$	$\int y \, dx$	indefinite integral
$\|z\|$	modulus of z	$\int_a^b y \, dx$	integral between the limits of a and b
$\arg z$	argument of z	$\oint y \, dx$	around a closed contour
x_i	ith value of the variate x	σ	standard deviation of a distributed variate
$\bar{x}$	average of several values of x		
ρ	correlation coefficient	s	standard deviation for a sample
r	correlation coefficient for a sample	n	number in a sample
p	probability	w	range
$\angle$	angle	$\therefore$	therefore
$\subset$	one subset	$\triangle$	triangle
$\mathscr{E}$	universal set	$\cup$	union
A^{-1}	inverse of the matrix A	$\cap$	intersection
$\parallel$	parallel to	A'	transpose of the matrix A
$\perp$	perpendicular to		

Multiples and Submultiples

Multiplying factor	Prefix	Symbol	Multiplying factor	Prefix	Symbol
10^{12}	tera	T	10^{-6}	micro	μ
10^{9}	giga	G	10^{-9}	nano	n
10^{6}	mega	M	10^{-12}	pico	p
10^{3}	kilo	k	10^{-15}	femto	f
10^{-3}	milli	m	10^{-18}	atto	a

Greek letters

A α alpha	E ϵ epsilon	I ι iota	N ν nu	P ρ rho	Φ ϕ phi
B β beta	Z ζ zeta	K κ kappa	Ξ ξ xi	Σ σ sigma	X χ chi
Γ γ gamma	H η eta	Λ λ lambda	O o omicron	T τ tau	Ψ ψ psi
Δ δ delta	Θ θ theta	M μ mu	Π π pi	Y υ upsilon	Ω ω omega

Standard symbols and units for physical quantities

Quantity	Symbol	Unit	Quantity	Symbol	Unit
Acceleration–gravitational	g	m/s²	Frequency	f	Hz
Acceleration–linear	a	m/s²	Frequency, resonant	f_r	Hz
Admittance	Y	S	Gravitational acceleration	g	m/s²
Altitude above sea level	z	m	Gibbs function	G	J
Amount of substance	n	mol	Gibbs function, specific	g	kJ/kg
Angle–plane	$\alpha, \beta, \theta, \phi$	rad			
Angle–solid	Ω, ω	steradian	Heat capacity, specific	c	kJ/kg K
Angular acceleration	α	rad/s²	Heat flow rate	ϕ	W
Angular velocity	ω	rad/s	Heat flux intensity	ϕ	kW/m²
Area	A	m²			
Area–second moment of	I	m⁴	Illumination	E	lx
			Impedance	Z	Ω
Bulk modulus	K	N/m², Pa	Inductance, self	L	H
			Inductance, mutual	M	H
Capacitance	C	μF	Internal energy	U, E	J
Capacity	V	ℓ	Internal energy, specific	u, e	kJ/kg
Coefficient of friction	μ	no unit	Inertia, moment of	I, J	kg m²
Coefficient of linear					
expansion	α	/°C	Kinematic viscosity	ν	m²/s, St
Conductance, electrical	G	S			
Conductance, thermal	h	kW/m²K	Length	l	m
Conductivity, electrical	σ	kS/mm	Light–velocity of	c	m/s
Conductivity, thermal	λ	W/m K	Light–wavelength of	λ	m
Cubical expansion–			Linear expansion–		
coefficient of	β	/°C	coefficient of	α	/°C
Current, electrical	I	A	Luminance	L	cd/m²
Current density	J	A/mm²	Luminous flux	ϕ	lm
			Luminous intensity	I	cd
Density	ρ	kg/m³			
Density, relative	d	no unit	Magnetic field strength	H	A/m
Dryness fraction	x	no unit	Magnetic flux	Φ	Wb
Dynamic viscosity	η	Ns/m², cP	Magnetic flux density	B	T
			Magnetomotive force	F	A
Efficiency	η	no unit	Mass, macroscopic	m	kg
Elasticity, modulus of	E	N/m², Pa	Mass, microscopic	M	u
Electric field strength	E	V/m	Mass, rate of flow	V	m³/s
Electric flux	ϕ	C	Mass, velocity	G	kg/m²s
Electric flux density	D	C/m²	Modulus, bulk	K	N/m²
Energy	W	J	Modulus of elasticity	E	N/m²
Energy, internal	U, E	J	Modulus of rigidity	G	N/m²
Energy, specific internal	u, e	kJ/kg	Modulus of section	Z	m³
Enthalpy	H	J	Molar mass of gas	M	kg/k mol
Enthalpy, specific	h	kJ/kg	Molar volume	V_m	m³/k mol
Entropy	S	kJ/K	Moment of force	M	Nm
Expansion–coefficient			Moment of inertia	I, J	kg m²
of cubical	β	/°C	Mutual inductance	M	H
Expansion–coefficient					
of linear	α	/°C	Number of turns in a		
			winding	N	no unit
Field strength, electric	E	V/m			
Field strength, magnetic	H	A/m	Periodic time	T	s
Flux density, electric	D	C/m²	Permeability, absolute	μ	μH/m
Flux density, magnetic	B	T	Permeability, absolute of		
Flux, electric	ψ	C	free space	μ_0	μH/m
Flux, magnetic	Φ	Wb	Permeability, relative	μ_r	
Force	F	N	Permeance	Λ	H
Force, resisting	R	N	Permittivity, absolute	ϵ	pF/m

2

Quantity	Symbol	Unit	Quantity	Symbol	Unit
Permittivity of free space	ϵ_0	pF/m	Stress, direct	σ	N/m², Pa
Permittivity, relative	ϵ_r	no unit	Shear modulus of		
Poisson's ratio	ν	no unit	rigidity	G	N/m², Pa
Polar moment of area	J	m⁴	Surface tension	γ	N/m
Power, apparent	S	VA	Susceptance	B	S
Power, active	P	W			
Power, reactive	Q	VA$_r$	Temperature value	θ	°C
Pressure	p	N/m², Pa	Temperature coefficients		
			of resistance	α, β, γ	/°C
Quantity of heat	Q	J	Thermodynamic		
Quantity of electricity	Q	Ah, C	temperature value	T	K
			Time	t	s
Reactance	X	Ω	Torque	T	Nm
Reluctance	S	/H, A/Wb			
Relative density	d	no unit	Vapour velocity	C	m/s
Resistance, electrical	R	Ω	Velocity	v	m/s
Resisting force	R	N	Velocity, angular	ω	rad/s
Resistance, temperature			Velocity of light	c	Mm/s
coefficients of	α, β, γ	/°C	Velocity of sound	a	m/s
Resistivity, conductors	ρ	M Ω mm	Voltage	V	V
Resistivity, insulators	ρ	M Ω mm	Volume	V	m³
Resonant frequency	f_r	Hz	Volume, rate of flow	V	m³/s
			Viscosity, dynamic	η	Ns/m², cP
Second moment of area	I	m⁴	Viscosity, kinematic	ν	m²/s, cSt
Self inductance	L	H			
Shear strain	γ	no unit	Wavelength	λ	m
Shear stress	τ	N/m², Pa	Work	W	J
Specific gas constant	R	kJ/kg K			
Specific heat capacity	c	kJ/kg K	Young's modulus of		
Specific volume	v	m³/kg	elasticity	E	N/m², Pa
Strain, direct	ϵ	no unit			

Abbreviations for units

Unit	abb.	Unit	abb.	Unit	abb.	Unit	abb.
metre	m	steradian	sr	newton	N	mole	mol
angström	A	radian per		bar	bar	watt	W
square metre	m²	second	rad/s	millibar	mb	decibel	dB
cubic metre	m³	hertz	Hz	standard		kelvin	K
litre	ℓ	revolution per		atmosphere	atm	centigrade	°C
second	s	minute	rev/min	millimetre of		coulomb	C
minute	min.	kilogramme	kg	mercury	mm Hg	ampere	A
hour	h	gramme	g	poise	P	volt	V
lumen	lm	tonne		stokes	S, St	ohm	Ω
candela	cd	(= 1 Mg)	t	joule	J	farad	F
lux	lx	seimen	S	kilowatt hour	kW h	henry	H
day	d	atomic mass		electron volt	eV	weber	Wb
year	a	unit	u	calorie	cal	tesla	T
radian	rad	pascal	Pa				

Chemical symbols and atomic weights

Element	Symbol	Atomic number	Atomic weight	Element	Symbol	Atomic number	Atomic weight
Actinium	Ac	89	(227)	Molybdenum	Mo	42	95.9
Aluminium	Al	13	26.9815	Neodymium	Nd	60	144.2
Americium	Am	95	(243)	Neon	Ne	10	20.179
Antimony	Sb	51	124.7	Neptunium	Np	93	237.0482
Argon	A	18	39.948	Nickel	Ni	28	58.7
Arsenic	As	33	74.9216	Niobium	Nb	41	92.9064
Astatine	At	85	~210	Nitrogen	N	7	14.0067
Barium	Ba	56	137.3	Nobelium	No	102	(254)
Berkelium	Bk	97	(247)	Osmium	Os	76	190.2
Beryllium	Be	4	9.01218	Oxygen	O	8	15.999
Bismuth	Bi	83	208.9806	Palladium	Pd	46	106.4
Boron	B	5	10.81	Phosphorus	P	15	30.9738
Bromine	Br	35	79.904	Platinum	Pt	78	195.0
Cadmium	Cd	48	112.40	Plutonium	Pu	94	(244)
Californium	Cf	98	(251)	Potassium	K	19	29.102
Calcium	Ca	20	40.08	Praseodymium	Pr	59	140.907
Carbon	C	6	12.011	Protoactinium	Pa	91	231.0359
Cerium	Ce	58	140.12	Polonium	Po	84	(210)
Cesium	Cs	55	132.9055	Promethium	Pm	61	(145)
Chlorine	Cl	17	35.453	Radium	Ra	88	226.0254
Chromium	Cr	24	51.996	Radon	Rn	86	(~222)
Cobalt	Co	27	58.9332	Rhenium	Re	75	186.2
Copper	Cu	29	63.546	Rhodium	Rh	45	102.9055
Curium	Cm	96	(247)	Rubidium	Rb	37	85.467
Dysprobium	Dy	66	162.50	Ruthenium	Ru	44	101.0
Erbium	Er	68	167.26	Samarium	Sm	62	150.4
Europium	Eu	63	151.96	Scandium	Sc	21	44.9559
Fermium	Fm	100	(257)	Selenium	Se	34	78.96
Fluorine	F	9	18.9984	Silicon	Si	14	28.086
Gadolinium	Gd	64	157.2	Silver	Ag	47	107.868
Gallium	Ga	31	69.72	Sodium	Na	11	22.9898
Germanium	Ge	32	72.59	Strontium	Sr	38	87.62
Gold	Au	79	196.9665	Sulphur	S	16	32.06
Hafnium	Hf	72	178.49	Tantalum	Ta	73	180.947
Helium	He	2	4.00260	Technetium	Tc	43	98.9062
Holmium	Ho	67	164.9303	Tellurium	Te	52	127.60
Hydrogen	H	1	1.0080	Terbium	Tb	65	158.9254
Indium	In	49	114.82	Thallium	Tl	81	204.37
Iodine	I	53	126.9045	Thorium	Th	90	232.0381
Iridium	Ir	77	193.2	Thulium	Tm	69	168.9342
Iron	Fe	26	55.84	Tin	Sn	50	118.6
Krypton	Kr	36	83.86	Titanium	Ti	22	49.9
Lanthanum	La	57	138.905	Tungsten	W	74	183.8
Lawrencium	Lr	103	(257)	Uranium	U	92	238.029
Lead	Pb	82	207.2	Vanadium	V	23	50.941·
Lithium	Li	3	6.941	Xenon	Xe	54	131.30
Lutetium	Lu	71	174.97	Ytterbium	Yb	70	173.0
Magnesium	Mg	12	24.305	Yttrium	Y	39	88.9059
Manganese	Mn	25	54.9380	Zinc	Zn	30	65.3
Mendelevium	Md	101	(256)	Zirconium	Zr	40	91.22
Mercury	Hg	80	200.5				

Specific heat of various substances

Substance	S.H. capacity kJ/kgK	Substance	S.H. capacity kJ/kgK	Substance	S.H. capacity kJ/kgK
Alcohol	2.604	Graphite	0.842	Quartz	0.787
Aluminium	0.896	Ice	2.110	Sand	0.816
Antimony	0.214	Iron, cast	0.544	Silica	0.800
Benzine	1.884	Iron, wrought	0.461	Silver	0.234
Brass	0.394	Kerosene	2.093	Soda	0.967
Brickwork	0.837	Lead	0.130	Steel, mild	0.486
Cadmium	0.239	Limestone	0.909	Steel, high carbon	0.490
Charcoal	0.837	Magnesia	0.930	Stone	0.837
Chalk	0.900	Marble	0.879	Sulphur	0.745
Coal	1.005	Masonry, brick	0.837	Sulphuric acid	1.382
Coke	0.850	Mercury	0.138	Tin	0.234
Copper	0.394	Naptha	1.298	Turpentine	1.976
Corundum	0.829	Nickel	0.456	Water	4.187
Ether	2.106	Oil, machine	1.675	Wood, fir	2.721
Fusel oil	2.361	Oil, olive	1.465	Wood, oak	2.387
Glass	0.812	Phosphorus	0.791	Wood, pine	1.955
Gold	0.130	Platinum	0.134	Zinc	0.398

Boiling points at atmospheric pressure

Substance	B.P. °C.	Substance	B.P. °C.	Substance	B.P. °C.
Alcohol	78	Ether	38	Brine	108
Alcohol wood	66	Linseed oil	264	Sulphuric acid	310
Ammonia	− 33	Mercury	358	Water, pure	100
Benzine	80	Napthalene	220	Water, sea	100.7
Bromine	63	Nitric acid	120		
Chloroform	60	Turpentine	157		

Loudness of sounds

Source	Intensity in decibels	Source	Intensity in decibels
Threshold of hearing	0	Loud conversation	70
Virtual silence	10	Door slamming	80
Quiet room	20	Riveting gun	90
Average home	30	Loud motor horn	100
Motor car	40	Thunder	110
Ordinary conversation	50	Aero-engine	120
Street traffic	60	Threshold of pain	130

Densities of various substances

Substance	Density g/cm³	Substance	Density g/cm³	Substance	Density g/cm³
Alcohol	0.79	Emery	4.0	Palm oil	0.97
Ammonia	0.89	Ether, sulphuric	0.72	Phosphorus	1.8
Asbestos	2.8	Fluoric acid	1.50	Petroleum oil	0.82
Benzine	0.69	Gasoline	0.70	Phosphoric acid	1.78
Borax	1.75	Glass	2.6	Quartz	2.6
Brick, common	1.8	Granite	2.65	Rape oil	0.92
Brick, fire	2.3	Gravel	1.75	Salt, common	2.1
Brick, hard	2.0	Gypsum	2.2	Sand, dry	1.6
Brick, pressed	2.15	Ice	0.9	Sand, wet	2.0
Brickwork in mortar	1.6	Ivory	1.85	Sandstone	2.3
Brickwork in cement	1.8	Kerosene	0.80	Slate	2.8
Cement	3.1	Limestone	2.6	Soapstone	2.7
Chalk	2.6	Linseed oil	0.92	Soil, black	2.0
Charcoal	0.4	Marble	2.7	Sulphur	2.0
Coal, anthracite	1.5	Masonry	2.4	Sulphuric acid	1.84
Coal, bituminious	1.27	Mica	2.8	Tar	1.00
Concrete	2.2	Mineral oil	0.92	Tile	1.8
Carbolic acid	0.96	Mortar	1.5	Turpentine	0.87
Carbon disulphide	1.26	Muriatic acid	1.2	Vinegar	1.08
Cotton seed oil	0.93	Naptha	0.76	Water	1.00
Earth, loose	1.2	Nitric acid	1.22	Water, sea	1.03
Earth, rammed	1.6	Olive oil	0.92	Whale oil	0.93

Latent heat of evaporation

Liquid	kJ/kg	Liquid	kJ/kg	Liquid	kJ/kg
Alcohol, ethyl	863	Bisulphide of		Sulphur dioxide	381
Alcohol, methyl	1119	carbon	372	Turpentine	309
Ammonia	1230	Ether	379	Water	2248

Latent heat of fusion

Substance	kJ/kg	Substance	kJ/kg	Substance	kJ/kg
Aluminium	387	Paraffin	147.2	Sulphur	39.2
Bismuth	52.9	Phosphorus	21.1	Tin	59.7
Cast iron, grey	96.3	Lead	23.3	Zinc	117.8
Cast iron, white	138.2	Silver	88.2	Ice	334.9
Copper	180	Nickel	309	Magnesium	372

Abbreviations for words

Word	abb.	Word	abb.	Word	abb.	Word	abb.
absolute	abs.	crystalline	cryst.	infra-red	i.r.	relative	
alternating		decompo-		magneto-		humidity	r.h.
current	a.c.	sition	decomp.	motive		root mean	
anhydrous	anhyd.	dilute	dil.	force	m.m.f.	square	r.m.s.
aqueous	aq.	direct current	d.c.	maximum	max.	temperature	temp.
boiling point	b.p.	electromotive		melting point	m.p.	standard	
calculated	calc.	force	e.m.f.	minimum	min.	temp. and	
concentrated	conc.	equation	eqn.	potential		pressure	s.t.p.
constant	const.	equivalent	equiv.	difference	p.d.	ultra violet	u.v.
corrected	corr.	experiment(al)	expt.	recrystallised	recryst.		
critical	crit.	freezing point	f.p.				

Commonly used constants

Constant	Numerical value	Logarithm	Constant	Numerical value	Logarithm
π	3.141593	0.4972	$1/\pi$	0.318310	$\bar{1}$.5029
2π	6.283185	0.7982	$\sqrt{\pi}$	1.772454	0.2486
$\pi/4$	0.785398	$\bar{1}$.8951	e	2.71828	0.4343
π^2	9.869604	0.9943	g	9.81	0.9917

Fixed points

Boiling point of liquid oxygen	$-182.97°C$
Melting point of ice (secondary point)	$0.00°C$
Triple point of water	$0.01°C$
Boiling point of water	$100°C$
Freezing point of zinc (secondary point)	$419.505°C$
Boiling point of liquid sulphur	$444.60°C$
Freezing point of liquid antimony	$630.50°C$
Melting point of silver	$960.80°C$
Melting point of gold	$1063.00°C$

Atmospheric pressure = 760 mm Hg = 1013 mb   1 bar = 10^2 kPa

Diameter of the earth = 12 750 km at the equator and 12 710 km at the poles.

Average radius of the earth = 6371 km

Speed of rotation of the earth = 1670 km/h

Calculator check

Not all calculators have the same logic and the keys and stores do not always work in the same way. Therefore before starting to perform strings of calculations it pays to check that the calculator is working correctly. The following can be used to check the logic of the calculator.

$67.84 + 91.92 + 71.85 = 231.61$

$66.32 - 19.85 = 46.47$

$88.56 - 13.84 + 24.31 = 99.03$

$77.3 \times 64.8 = 5009.04$

$91.76 \times 3.84 + 817.52 = 1169.8784$

$(7.85 + 3.91) \times 83.64 = 983.6064$

$91.3 \times 43.2 \times 68.0 = 268\ 202.88$

$\dfrac{91.76}{1.85} = 49.6$

$\dfrac{81.32 \times 14.63}{76.51} = 15.549\ 752$

$\dfrac{84.3}{91.2} + \dfrac{76.51}{3.84} = 20.848\ 821$

for calculators without a memory this may be calculated thus

$\left(\dfrac{84.3 \times 3.84}{91.2} + 76.51\right) \div 3.84 = 20.848\ 821$

$\dfrac{816.1}{94.3} - \dfrac{36.2}{14.7} = 6.191\ 7098$

for calculators without a memory this may be calculated thus

$\left(\dfrac{-36.2 \times 94.3}{14.7} + 816.1\right) \div 94.3 = 6.191\ 7098$

$17.62 - \dfrac{8.54}{3.61} = 15.254\ 35$

Factors

$$(a+b)^2 = a^2 + 2ab + b^2$$
$$(a-b)^2 = a^2 - 2ab + b^2$$
$$a^3 + b^3 = (a+b)(a^2 - ab + b^2)$$
$$a^3 - b^3 = (a-b)(a^2 + ab + b^2)$$
$$a^2 - b^2 = (a+b)(a-b)$$
$$(a+b)^3 = a^3 + 3a^2b + 3ab^2 + b^3$$
$$(a-b)^3 = a^3 - 3a^2b + 3ab^2 - b^3$$

Indices

$$a^m \times a^n = a^{m+n}$$
$$a^m \div a^n = a^{m-n}$$
$$(a^m)^n = a^{mn}$$
$$\sqrt[m]{a^n} = a^{m/n}$$
$$\frac{1}{a^n} = a^{-n}$$
$$a^0 = 1$$

Logarithms

If $N = a^x$ then $\log_a N = x$ and $N = a^{\log_a N}$

$$\log_a N = \frac{\log_b N}{\log_b a}$$
$$\log(ab) = \log a + \log b$$
$$\log\left(\frac{a}{b}\right) = \log a - \log b$$
$$\log a^n = n \log a$$
$$\log \sqrt[n]{a} = \frac{1}{n}\log a$$
$$\log_a 1 = 0$$
$$\log_e N = 2.3026 \log_{10} N$$

Quadratic equation

If $ax^2 + bx + c = 0$
$$x = \frac{-b \pm \sqrt{b^2 - 4ac}}{2a}$$

Graphs

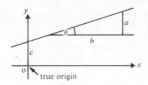

$$m = \tan \theta = \frac{a}{b}$$

The equation of a straight line can be written in the form $y = mx + c$ where m is the gradient of the line and c is the intercept on the y-axis.
Non-linear relationships can sometimes be converted into linear relationships. The most common of these are given in the table below:

Equation	Plot	Gradient	Intercept
$y = ax^n + b$	$y \text{ v } x^n$	a	b
$y = \dfrac{a}{x^n} + b$	$y \text{ v } \dfrac{1}{x^n}$	a	b
$y = a\sqrt[n]{x} + b$	$y \text{ v } \sqrt[n]{x}$	a	b
$y = ax^n + bx^{n-1}$	$\dfrac{y}{x^{n-1}} \text{ v } x$	a	b
$y = ax^n$	$\log y \text{ v } \log x$	n	$\log a$
$y = ab^x$	$\log y \text{ v } x$	$\log b$	$\log a$
$y = ae^{bx}$	$\log y \text{ v } x$	$b \log e$	$\log a$

Variation

If $y \propto x$ then $y = kx$. This is direct variation.
If $y \propto \dfrac{1}{x}$ then $y = \dfrac{k}{x}$. This is inverse variation.

If p varies directly as t and inversely as v then $p = \dfrac{kt}{v}$.
This is joint variation.

Binomial theorem

$$(a+b)^n = a^n + na^{n-1}b + \frac{n(n-1)\,a^{n-2}\,b^2}{2!} + \dots b^n$$

$$(1+x)^n = 1 + nx + \frac{n(n-1)x^2}{2!} + \dots x^n$$

If n is a positive integer the series is finite and is true for all values of x.
If n is negative or fractional the series is infinite and is valid only if x lies between -1 and $+1$.

Series

$$e = 1 + 1 + \frac{1}{2!} + \frac{1}{3!} + \dots$$

$$e^x = 1 + x + \frac{x^2}{2!} + \frac{x^3}{3!} + \dots$$

$$e^{-x} = 1 - x + \frac{x^2}{2!} - \frac{x^3}{3!} + \dots$$

9

Areas of plane figures

Rectangle

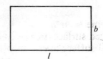

$\text{Area} = lb$
$\text{Perimeter} = 2l + 2b$

Parallelogram

$\text{Area} = bh$

Triangle

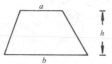

$$\begin{aligned}\text{Area} &= \tfrac{1}{2}bh = \sqrt{s(s-a)(s-b)(s-c)} \\ &= \tfrac{1}{2}ab \sin C\end{aligned}$$
Where $s = \dfrac{a+b+c}{2}$

Trapezium

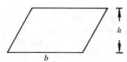

$\text{Area} = \tfrac{1}{2}h(a+b)$

Circle

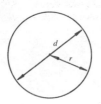

$\text{Area} = \pi r^2 = \dfrac{\pi d^2}{4}$
$\text{Circumference} = \pi d = 2\pi r$

Segment of a circle

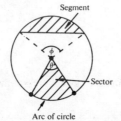

$\text{Area} = \tfrac{1}{2}r^2(\phi - \sin \phi)$
(ϕ in radians)

Sector of a circle

$\text{Area} = \pi r^2 \times \dfrac{\theta}{360}$

$\text{Length of arc} = 2\pi r \times \dfrac{\theta}{360}$
(θ in degrees)

Volumes and surface areas

Cylinder

Volume $= \pi r^2 h$
Curved surface area $= 2\pi rh$
Total surface area $= 2\pi rh + 2\pi r^2$
$ = 2\pi r(r+h)$

Any solid having a uniform cross-section

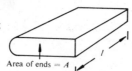

Area of ends $= A$

Volume $= Al$
Curved surface area
$ =$ perimeter of cross-section $\times$ length
Total surface area
$ =$ curved surface area $+$ area of ends

Cone

Volume $= \frac{1}{3}\pi r^2 h$ $\qquad$ ($h =$ vertical height)
Curved surface area $= \pi rl$ $\qquad$ ($l =$ slant height)
Total surface area $= \pi rl + \pi r^2$

Sphere

Volume $= \frac{4}{3}\pi r^3$
Surface area $= 4\pi r^2$

Frustrum of a cone

Volume $= \frac{1}{3}\pi h(R^2 + Rr + r^2)$
Curved surface area $= \pi(R+r)l$
Total surface area $= \pi(R+r)l + \pi R^2 + \pi r^2$

Pyramid

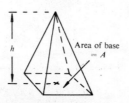

Area of base $= A$

Volume $= \frac{1}{3}Ah$

Prism

Any solid with two faces parallel and having a constant cross-section. The end faces must be triangles, quadrilaterals or polygons.

Volume $=$ area of cross-section $\times$ length of prism

Angles

$$1 \text{ revolution} = 360° = 2\pi \text{ radians}$$
$$60' = 1°$$
$$60'' = 1'$$

$$1° = \frac{2\pi}{360} \text{ radians}$$

$$1 \text{ radian} = \frac{360}{2\pi} = 57.3°$$

$$45° = \frac{\pi}{4} \text{ radians} \qquad 90° = \frac{\pi}{2} \text{ radians}$$

$$60° = \frac{\pi}{3} \text{ radians} \qquad 180° = \pi \text{ radians}$$

$$120° = \frac{2\pi}{3} \text{ radians} \qquad 270° = \frac{3\pi}{2} \text{ radians}$$

Acute angle
(less than 90°)

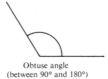

Obtuse angle
(between 90° and 180°)

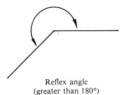

Reflex angle
(greater than 180°)

Complementary   angles are angles whose sum is 90°

Supplementary   angles are angles whose sum is 180°

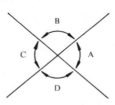

The vertically opposite angles are equal
$$\angle A = \angle C \text{ and } \angle B = \angle D$$

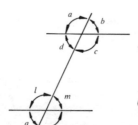

When two parallel lines are cut by a transversal
(i)   The corresponding angles are equal:
$$a = l; b = m; c = p; b = q$$
(ii)   The alternate angles are equal: $d = m; c = l$
(iii)   The interior angles are supplementary:
$$d + l = 180°; c + m = 180°$$

Triangles

Acute-angled (all angles less than 90°)

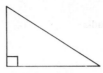

Right-angled (one angle = 90°)

Obtuse angled (one angle greater than 90°)

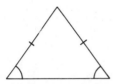

Isosceles triangle (two sides and two angles equal)

Equilateral triangle (all sides and all angles equal)

The sum of the angles of a triangle equals 180°

Pythagoras' theorem

In a right-angled triangle the square on the hypotenuse is equal to the sum of the squares on the other two sides.

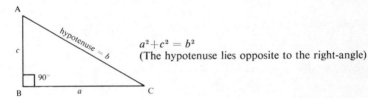

$$a^2 + c^2 = b^2$$
(The hypotenuse lies opposite to the right-angle)

Congruency

Two triangles are congruent if they are equal in every respect. Any of the following are sufficient to prove that two triangles are congruent:

(i) One side and two angles in one triangle equal to one side and two similarly located angles in the second triangle.

(ii) Two sides and the angle between them in one triangle equal to two sides and the angle between them in the second triangle.

(iii) Three sides of one triangle equal to three sides in the second triangle.

(iv) In right-angled triangles the hypotenuses are equal and one other side in each triangle also equal.

Similar triangles

Two triangles are similar if they are equi-angular. If in $\triangle$s ABC and XYZ, $\angle A = \angle X$, $\angle B = \angle Y$ and $\angle C = \angle Z$ then

$$\frac{AB}{XY} = \frac{AC}{XZ} = \frac{BC}{YZ}$$

Any of the following is sufficient to prove that two triangles are similar:

(i) Two angles in one triangle equal to two angles in the second triangle.

(ii) Two sides in one triangle are proportional to two sides in the second triangle and the angle between these sides in each triangle is equal.

(iii) Three sides in one triangle are proportional to the three sides in the second triangle.

13

Geometry of the circle

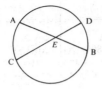

$$AE \cdot EB = CE \cdot ED$$

PQ = diameter (d)
RS $\perp$ PQ
PX = h
RX = XS = s

$$h(d-h) = x^2$$

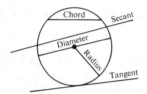

The angle which an arc of a circle subtends at the centre of a circle is twice the angle which the arc subtends at the circumference.

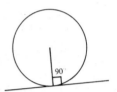

The tangent to a circle is at right-angles to a radius drawn from the point of tangency.

Trigonometry

$$\sin A = \frac{\text{opposite side}}{\text{hypotenuse}} = \frac{a}{b}$$

$$\cos A = \frac{\text{adjacent side}}{\text{hypotenuse}} = \frac{c}{b}$$

$$\tan A = \frac{\text{opposite side}}{\text{adjacent side}} = \frac{a}{c}$$

$$\text{cosec } A = \frac{1}{\sin A} = \frac{\text{hypotenuse}}{\text{opposite side}} = \frac{b}{a}$$

$$\sec A = \frac{1}{\cos A} = \frac{\text{hypotenuse}}{\text{adjacent side}} = \frac{b}{c}$$

$$\cot A = \frac{1}{\tan A} = \frac{\text{adjacent side}}{\text{opposite side}} = \frac{c}{a}$$

$$\sin 60° = \frac{\sqrt{3}}{2} \quad \sin 30° = \frac{1}{2} \quad \sin 45° = \frac{\sqrt{2}}{2}$$

$$\cos 60° = \frac{1}{2} \quad \cos 30° = \frac{\sqrt{3}}{2} \quad \cos 45° = \frac{\sqrt{2}}{2}$$

$$\tan 60° = \sqrt{3} \quad \tan 30° = \frac{\sqrt{3}}{3} \quad \tan 45° = 1$$

$$\cos A = \sin(90°-A)$$
$$\sin A = \cos(90°-A)$$

Trigonometrical identities

$$\sin^2 A + \cos^2 A = 1 \qquad \sec^2 A = 1 + \tan^2 A$$

$$\text{cosec}^2 A = 1 + \cot^2 A \qquad \tan A = \frac{\sin A}{\cos A}$$

The general angle

Quadrant	Angle	$\sin A =$	$\cos A =$	$\tan A =$
first	0 to 90°	$\sin A$	$\cos A$	$\tan A$
second	90° to 180°	$\sin(180°-A)$	$-\cos(180°-A)$	$-\tan(180°-A)$
third	180° to 270°	$-\sin(A-180°)$	$-\cos(A-180°)$	$\tan(A-180°)$
fourth	270° to 360°	$-\sin(360°-A)$	$\cos(360°-A)$	$-\tan(360°-A)$

For any triangle

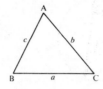

Sine rule
$$\frac{a}{\sin A} = \frac{b}{\sin B} = \frac{c}{\sin C}$$

Cosine rule
$$a^2 = b^2 + c^2 - 2bc\cos A$$
$$b^2 = a^2 + c^2 - 2ac\cos B$$
$$c^2 = a^2 + b^2 - 2ab\cos C$$

Tangent rule
$$\tan\frac{B-C}{2} = \frac{b-c}{b+c}\cot\frac{A}{2}$$

Half-angle formulae
$$\sin\frac{A}{2} = \sqrt{\frac{(s-b)(s-c)}{bc}}$$
$$\cos\frac{A}{2} = \sqrt{\frac{s(s-a)}{bc}}$$
$$\tan\frac{A}{2} = \sqrt{\frac{(s-b)(s-c)}{s(s-a)}}$$
where $s = \frac{1}{2}(a+b+c)$

Factor formulae
$$\cos P + \cos Q = 2\cos\frac{P+Q}{2}\cos\frac{P-Q}{2}$$
$$\cos P - \cos Q = -2\sin\frac{P+Q}{2}\sin\frac{P-Q}{2}$$
$$\sin P + \sin Q = 2\sin\frac{P+Q}{2}\cos\frac{P-Q}{2}$$
$$\sin P - \sin Q = 2\cos\frac{P+Q}{2}\sin\frac{P-Q}{2}$$

Multiple angles
$$\sin(A\pm B) = \sin A\cos B \pm \cos A\sin B$$
$$\cos(A\pm B) = \cos A\cos B \mp \sin A\sin B$$
$$\tan(A\pm B) = \frac{\tan A \pm \tan B}{1 \mp \tan A\tan B}$$
$$R\sin(\omega t\pm a) = a\sin\omega t \pm b\cos\omega t$$
where $R = \sqrt{a^2+b^2}$ and $\tan a = \frac{b}{a}$

Calculus

Derivatives		Integrals	
y	dy/dx	y	$\int y\,dx$
ax^n	anx^{n-1}	ax^n	$\dfrac{ax^{n+1}}{n+1}\quad n \neq -1$
$\sin ax$	$a\cos ax$	$\sin ax$	$-\dfrac{1}{a}\cos ax$
$\cos ax$	$-a\sin ax$	$\cos ax$	$\dfrac{1}{a}\sin ax$
$\tan ax$	$a\sec^2 ax$	e^{ax}	$\dfrac{1}{a}e^{ax}$
$\log_e ax$	$\dfrac{1}{x}$	$\dfrac{1}{ax}$	$\dfrac{1}{a}\log_e x$
e^{ax}	ae^{ax}		

Differentiation

Product rule: If $y = uv$, $\dfrac{dy}{dx} = v\dfrac{du}{dx} + u\dfrac{dv}{dx}$ u and v are both functions of x

Quotient rule: If $y = \dfrac{u}{v}$, $\dfrac{dy}{dx} = \dfrac{v\dfrac{du}{dx} - u\dfrac{dv}{dx}}{v^2}$

Function of a function: $\dfrac{dy}{dx} = \dfrac{dy}{du} \cdot \dfrac{du}{dx}$

Conditions for maximum and minimum:

y has a maximum value if $\dfrac{dy}{dx} = 0$ and $\dfrac{d^2y}{dx^2}$ is negative

y has a minimum value if $\dfrac{dy}{dx} = 0$ and $\dfrac{d^2y}{dx^2}$ is positive

Point of inflexion: $\dfrac{d^2y}{dx^2}$ is zero and changes sign

Integration

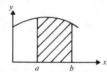

Area $A = \displaystyle\int_a^b y\,dx$

Volume generated by rotating the area A completely about the x-axis is

$\pi \displaystyle\int_a^b y^2\,dx.$

Centroid of a plane area

$$\bar{x} = \frac{\displaystyle\int_a^b xy\,dx}{\text{area}}$$

$$\bar{y} = \frac{\tfrac{1}{2}\displaystyle\int_a^b y^2\,dx}{\text{area}}$$

Second moment of area

$$I \text{ (about } y\text{-axis)} = \int_a^b x^2 y \, dx$$

$I = Ak^2$ where A is the area and k is the radius of gyration

Theorem of perpendicular axes

If OX and OY are two axes in the plane of a lamina and OZ is mutually perpendicular to OX and OY then

$$I_{OZ} = I_{OX} + I_{OY}$$

Theorem of parallel axes

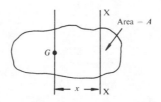

$$I_{XX} = I_G + Ax^2$$

Pappus' Theorem

Surface area of revolution $= 2\pi \bar{y} L$
Volume of revolution $\quad = 2\pi \bar{y} A$

Irregular plane areas

Mid-ordinate rule: Area $= b(h_1 + h_2 + h_3 + \dots h_n)$ see Fig. 1

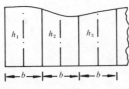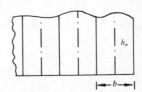

Fig. 1

Trapezium (Trapezoidal) rule: Area $= \dfrac{b}{2}\Big[(h_0 + h_n) + 2(h_1 + h_2 + \dots h_{n-1})\Big]$

see Fig. 2

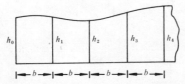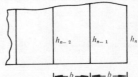

Fig. 2

Simpson's rule: Area $= \dfrac{b}{3}\Big[(h_0 + h_n) + 4(h_1 + h_3 + \dots h_{n-1}) + 2(h_2 + h_4 + \dots h_{n-2})\Big]$

(n must be even)

Centroids of some plane figures

Rectangle

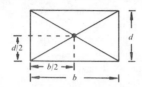

Centroid lies at the intersection of the diagonals.

Triangle

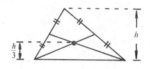

Centroid lies at the intersection of the bisectors of the sides.

Semi-circle

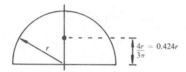

$\dfrac{4r}{3\pi} = 0.424r$

Quadrant of a circle

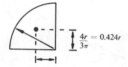

$\dfrac{4r}{3\pi} = 0.424r$

Segment of a circle

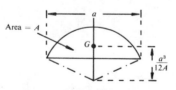

Area = A

$\dfrac{a^3}{12A}$

Sector of a circle

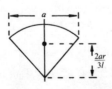

$\dfrac{2ar}{3l}$

l = length of arc

Properties of sections

Section	Section Modulus	Second Moment of Area

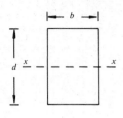

$$Z_{xx} = \frac{bd^2}{6}$$

$$I_{xx} = \frac{bd^3}{12}$$

$$Z_{xx} = \frac{BD^3 - bd^3}{6D}$$

$$I_{xx} = \frac{BD^3 - bd^3}{12}$$

$$Z_{xx} = \frac{\pi D^3}{32}$$

$$I_{xx} = \frac{\pi D^4}{64}$$

$$I_{\text{polar}} = \frac{\pi D^4}{32}$$

$$Z_{xx} = \frac{\pi}{32}\left(\frac{D^4 - d^4}{D}\right)$$

$$I_{xx} = \frac{\pi}{64}(D^4 - d^4)$$

$$I_{\text{polar}} = \frac{\pi}{32}(D^4 - d^4)$$

Logarithms

	0	1	2	3	4	5	6	7	8	9	1	2	3	4	5	6	7	8	9
10	0000	0043	0086	0128	0170						4	8	13	17	21	25	30	34	38
						0212	0253	0294	0334	0374	4	8	12	16	20	24	28	32	36
11	0414	0453	0492	0531	0569						4	8	12	15	19	23	27	31	35
						0607	0645	0682	0719	0755	4	7	11	15	18	22	26	30	33
12	0792	0828	0864	0899	0934						4	7	11	14	18	21	25	28	32
						0969	1004	1038	1072	1106	3	7	10	14	17	20	24	27	31
13	1139	1173	1206	1239	1271						3	7	10	13	16	20	23	26	30
						1303	1335	1367	1399	1430	3	6	9	13	16	19	22	25	28
14	1461	1492	1523	1553	1584						3	6	9	12	15	18	21	24	27
						1614	1644	1673	1703	1732	3	6	9	12	15	18	21	24	27
15	1761	1790	1818	1847	1875						3	6	9	11	14	17	20	23	26
						1903	1931	1959	1987	2014	3	6	8	11	14	17	19	22	25
16	2041	2068	2095	2122	2148						3	5	8	11	13	16	19	21	24
						2175	2201	2227	2253	2279	3	5	8	10	13	16	18	21	23
17	2304	2330	2355	2380	2405						3	5	8	10	13	15	18	20	23
						2430	2455	2480	2504	2529	2	5	7	10	12	15	17	20	22
18	2553	2577	2601	2625	2648						2	5	7	10	12	14	17	19	21
						2672	2695	2718	2742	2765	2	5	7	9	12	14	16	19	21
19	2788	2810	2833	2856	2878						2	5	7	9	11	14	16	18	20
						2900	2923	2945	2967	2989	2	4	7	9	11	13	15	18	20
20	3010	3032	3054	3075	3096	3118	3139	3160	3181	3201	2	4	6	8	11	13	15	17	19
21	3222	3243	3263	3284	3304	3324	3345	3365	3385	3404	2	4	6	8	10	12	14	16	18
22	3424	3444	3464	3483	3502	3522	3541	3560	3579	3598	2	4	6	8	10	12	14	15	17
23	3617	3636	3655	3674	3692	3711	3729	3747	3766	3784	2	4	6	7	9	11	13	15	17
24	3802	3820	3838	3856	3874	3892	3909	3927	3945	3962	2	4	5	7	9	11	12	14	16
25	3979	3997	4014	4031	4048	4065	4082	4099	4116	4133	2	3	5	7	9	10	12	14	15
26	4150	4166	4183	4200	4216	4232	4249	4265	4281	4298	2	3	5	7	8	10	11	13	15
27	4314	4330	4346	4362	4378	4393	4409	4425	4440	4456	2	3	5	6	8	9	11	13	14
28	4472	4487	4502	4518	4533	4548	4564	4579	4594	4609	2	3	5	6	8	9	11	12	14
29	4624	4639	4654	4669	4683	4698	4713	4728	4742	4757	1	3	4	6	7	9	10	12	13
30	4771	4786	4800	4814	4829	4843	4857	4871	4886	4900	1	3	4	6	7	9	10	11	13
31	4914	4928	4942	4955	4969	4983	4997	5011	5024	5038	1	3	4	6	7	8	10	11	12
32	5051	5065	5079	5092	5105	5119	5132	5145	5159	5172	1	3	4	5	7	8	9	11	12
33	5185	5198	5211	5224	5237	5250	5263	5276	5289	5302	1	3	4	5	6	8	9	10	12
34	5315	5328	5340	5353	5366	5378	5391	5403	5416	5428	1	3	4	5	6	8	9	10	11
35	5441	5453	5465	5478	5490	5502	5514	5527	5539	5551	1	2	4	5	6	7	9	10	11
36	5563	5575	5587	5599	5611	5623	5635	5647	5658	5670	1	2	4	5	6	7	8	10	11
37	5682	5694	5705	5717	5729	5740	5752	5763	5775	5786	1	2	3	5	6	7	8	9	10
38	5798	5809	5821	5832	5843	5855	5866	5877	5888	5899	1	2	3	5	6	7	8	9	10
39	5911	5922	5933	5944	5955	5966	5977	5988	5999	6010	1	2	3	4	5	7	8	9	10
40	6021	6031	6042	6053	6064	6075	6085	6096	6107	6117	1	2	3	4	5	6	8	9	10
41	6128	6138	6149	6160	6170	6180	6191	6201	6212	6222	1	2	3	4	5	6	7	8	9
42	6232	6243	6253	6263	6274	6284	6294	6304	6314	6325	1	2	3	4	5	6	7	8	9
43	6335	6345	6355	6365	6375	6385	6395	6405	6415	6425	1	2	3	4	5	6	7	8	9
44	6435	6444	6454	6464	6474	6484	6493	6503	6513	6522	1	2	3	4	5	6	7	8	9
45	6532	6542	6551	6561	6571	6580	6590	6599	6609	6618	1	2	3	4	5	6	7	8	9
46	6628	6637	6646	6656	6665	6675	6684	6693	6702	6712	1	2	3	4	5	6	7	7	8
47	6721	6730	6739	6749	6758	6767	6776	6785	6794	6803	1	2	3	4	5	5	6	7	8
48	6812	6821	6830	6839	6848	6857	6866	6875	6884	6893	1	2	3	4	4	5	6	7	8
49	6902	6911	6920	6928	6937	6946	6955	6964	6972	6981	1	2	3	4	4	5	6	7	8

$$\log(3.546 \times 8.714) = \log 3.546 + \log 8.714$$
$$= 0.5497 + 0.9402 = 1.4899$$
by taking antilog of 1.4899
$$3.546 \times 8.714 = 30.89$$

$$\log\left(\frac{7.362}{5.169}\right) = \log 7.362 - \log 5.169$$
$$= 0.8670 - 0.7134 = 0.1536$$
by taking antilog of 0.1536
$$\left(\frac{7.362}{5.169}\right) = 1.424$$

The characteristic for a number greater than 1 is found by subtracting 1 from the number of figures to the left of the decimal point.

$$\log 384.2 = 2.5845 \qquad \log 59\,860 = 4.7771$$

Logarithms

	0	1	2	3	4	5	6	7	8	9	1	2	3	4	5	6	7	8	9
50	6990	6998	7007	7016	7024	7033	7042	7050	7059	7067	1	2	3	3	4	5	6	7	8
51	7076	7084	7093	7101	7110	7118	7126	7135	7143	7152	1	2	3	3	4	5	6	7	8
52	7160	7168	7177	7185	7193	7202	7210	7218	7226	7235	1	2	2	3	4	5	6	7	7
53	7243	7251	7259	7267	7275	7284	7292	7300	7308	7316	1	2	2	3	4	5	6	6	7
54	7324	7332	7340	7348	7356	7364	7372	7380	7388	7396	1	2	2	3	4	5	6	6	7
55	7404	7412	7419	7427	7435	7443	7451	7459	7466	7474	1	2	2	3	4	5	5	6	7
56	7482	7490	7497	7505	7513	7520	7528	7536	7543	7551	1	2	2	3	4	5	5	6	7
57	7559	7566	7574	7582	7589	7597	7604	7612	7619	7627	1	2	2	3	4	5	5	6	7
58	7634	7642	7649	7657	7664	7672	7679	7686	7694	7701	1	1	2	3	4	4	5	6	7
59	7709	7716	7723	7731	7738	7745	7752	7760	7767	7774	1	1	2	3	4	4	5	6	7
60	7782	7789	7796	7803	7810	7818	7825	7832	7839	7846	1	1	2	3	4	4	5	6	6
61	7853	7860	7868	7875	7882	7889	7896	7903	7910	7917	1	1	2	3	4	4	5	6	6
62	7924	7931	7938	7945	7952	7959	7966	7973	7980	7987	1	1	2	3	3	4	5	6	6
63	7993	8000	8007	8014	8021	8028	8035	8041	8048	8055	1	1	2	3	3	4	5	5	6
64	8062	8069	8075	8082	8089	8096	8102	8109	8116	8122	1	1	2	3	3	4	5	5	6
65	8129	8136	8142	8149	8156	8162	8169	8176	8182	8189	1	1	2	3	3	4	5	5	6
66	8195	8202	8209	8215	8222	8228	8235	8241	8248	8254	1	1	2	3	3	4	5	5	6
67	8261	8267	8274	8280	8287	8293	8299	8306	8312	8319	1	1	2	3	3	4	5	5	6
68	8325	8331	8338	8344	8351	8357	8363	8370	8376	8382	1	1	2	3	3	4	4	5	6
69	8388	8395	8401	8407	8414	8420	8426	8432	8439	8445	1	1	2	2	3	4	4	5	6
70	8451	8457	8463	8470	8476	8482	8488	8494	8500	8506	1	1	2	2	3	4	4	5	6
71	8513	8519	8525	8531	8537	8543	8549	8555	8561	8567	1	1	2	2	3	4	4	5	5
72	8573	8579	8585	8591	8597	8603	8609	8615	8621	8627	1	1	2	2	3	4	4	5	5
73	8633	8639	8645	8651	8657	8663	8669	8675	8681	8686	1	1	2	2	3	4	4	5	5
74	8692	8698	8704	8710	8716	8722	8727	8733	8739	8745	1	1	2	2	3	3	4	5	5
75	8751	8756	8762	8768	8774	8779	8785	8791	8797	8802	1	1	2	2	3	3	4	5	5
76	8808	8814	8820	8825	8831	8837	8842	8848	8854	8859	1	1	2	2	3	3	4	5	5
77	8865	8871	8876	8882	8887	8893	8899	8904	8910	8915	1	1	2	2	3	3	4	4	5
78	8921	8927	8932	8938	8943	8949	8954	8960	8965	8971	1	1	2	2	3	3	4	4	5
79	8976	8982	8987	8993	8998	9004	9009	9015	9020	9025	1	1	2	2	3	3	4	4	5
80	9031	9036	9042	9047	9053	9058	9063	9069	9074	9079	1	1	2	2	3	3	4	4	5
81	9085	9090	9096	9101	9106	9112	9117	9122	9128	9133	1	1	2	2	3	3	4	4	5
82	9138	9143	9149	9154	9159	9165	9170	9175	9180	9186	1	1	2	2	3	3	4	4	5
83	9191	9196	9201	9206	9212	9217	9222	9227	9232	9238	1	1	2	2	3	3	4	4	5
84	9243	9248	9253	9258	9263	9269	9274	9279	9284	9289	1	1	2	2	3	3	4	4	5
85	9294	9299	9304	9309	9315	9320	9325	9330	9335	9340	1	1	2	2	3	3	4	4	5
86	9345	9350	9355	9360	9365	9370	9375	9380	9385	9390	1	1	2	2	3	3	4	4	5
87	9395	9400	9405	9410	9415	9420	9425	9430	9435	9440	0	1	1	2	2	3	3	4	4
88	9445	9450	9455	9460	9465	9469	9474	9479	9484	9489	0	1	1	2	2	3	3	4	4
89	9494	9499	9504	9509	9513	9518	9523	9528	9533	9538	0	1	1	2	2	3	3	4	4
90	9542	9547	9552	9557	9562	9566	9571	9576	9581	9586	0	1	1	2	2	3	3	4	4
91	9590	9595	9600	9605	9609	9614	9619	9624	9628	9633	0	1	1	2	2	3	3	4	4
92	9638	9643	9647	9652	9657	9661	9666	9671	9675	9680	0	1	1	2	2	3	3	4	4
93	9685	9689	9694	9699	9703	9708	9713	9717	9722	9727	0	1	1	2	2	3	3	4	4
94	9731	9736	9741	9745	9750	9754	9759	9763	9768	9773	0	1	1	2	2	3	3	4	4
95	9777	9782	9786	9791	9795	9800	9805	9809	9814	9818	0	1	1	2	2	3	3	4	4
96	9823	9827	9832	9836	9841	9845	9850	9854	9859	9863	0	1	1	2	2	3	3	4	4
97	9868	9872	9877	9881	9886	9890	9894	9899	9903	9908	0	1	1	2	2	3	3	4	4
98	9912	9917	9921	9926	9930	9934	9939	9943	9948	9952	0	1	1	2	2	3	3	4	4
99	9956	9961	9965	9969	9974	9978	9983	9987	9991	9996	0	1	1	2	2	3	3	3	4

$$\log(2.534)^3 = 3 \times \log 2.534$$
$$= 3 \times 0.4038 = 1.2114$$
by taking antilog of 1.2114
$$(2.534)^3 = 16.28$$

$$\log \sqrt[5]{8.176} = \tfrac{1}{5} \times \log 8.176$$
$$= \tfrac{1}{5} \times 0.9125 = 0.1825$$
by taking antilog of 0.1825
$$\sqrt[5]{8.176} = 1.523$$

The negative characteristic for a number between 0 and 1 is found by adding 1 to the number of zeros following the decimal point.

$$\log 0.0578 = \overline{2}.7619 \qquad \log 0.000\,432 = \overline{4}.6355$$

Antilogarithms

	0	1	2	3	4	5	6	7	8	9	1	2	3	4	5	6	7	8	9
0.00	1000	1002	1005	1007	1009	1012	1014	1016	1019	1021	0	0	1	1	1	1	2	2	2
0.01	1023	1026	1028	1030	1033	1035	1038	1040	1042	1045	0	0	1	1	1	1	2	2	2
0.02	1047	1050	1052	1054	1057	1059	1062	1064	1067	1069	0	0	1	1	1	1	2	2	2
0.03	1072	1074	1076	1079	1081	1084	1086	1089	1091	1094	0	0	1	1	1	1	2	2	2
0.04	1096	1099	1102	1104	1107	1109	1112	1114	1117	1119	0	1	1	1	1	2	2	2	2
0.05	1122	1125	1127	1130	1132	1135	1138	1140	1143	1146	0	1	1	1	1	2	2	2	2
0.06	1148	1151	1153	1156	1159	1161	1164	1167	1169	1172	0	1	1	1	1	2	2	2	2
0.07	1175	1178	1180	1183	1186	1189	1191	1194	1197	1199	0	1	1	1	1	2	2	2	2
0.08	1202	1205	1208	1211	1213	1216	1219	1222	1225	1227	0	1	1	1	1	2	2	2	3
0.09	1230	1233	1236	1239	1242	1245	1247	1250	1253	1256	0	1	1	1	1	2	2	2	3
0.10	1259	1262	1265	1268	1271	1274	1276	1279	1282	1285	0	1	1	1	1	2	2	2	3
0.11	1288	1291	1294	1297	1300	1303	1306	1309	1312	1315	0	1	1	1	2	2	2	2	3
0.12	1318	1321	1324	1327	1330	1334	1337	1340	1343	1346	0	1	1	1	2	2	2	3	3
0.13	1349	1352	1355	1358	1361	1365	1368	1371	1374	1377	0	1	1	1	2	2	2	3	3
0.14	1380	1384	1387	1390	1393	1396	1400	1403	1406	1409	0	1	1	1	2	2	3	3	3
0.15	1413	1416	1419	1422	1426	1429	1432	1435	1439	1442	0	1	1	1	2	2	2	3	3
0.16	1445	1449	1452	1455	1459	1462	1466	1469	1472	1476	0	1	1	1	2	2	2	3	3
0.17	1479	1483	1486	1489	1493	1496	1500	1503	1507	1510	0	1	1	1	2	2	2	3	3
0.18	1514	1517	1521	1524	1528	1531	1535	1538	1542	1545	0	1	1	1	2	2	2	3	3
0.19	1549	1552	1556	1560	1563	1567	1570	1574	1578	1581	0	1	1	1	2	2	3	3	3
0.20	1585	1589	1592	1596	1600	1603	1607	1611	1614	1618	0	1	1	1	2	2	3	3	3
0.21	1622	1626	1629	1633	1637	1641	1644	1648	1652	1656	0	1	1	2	2	2	3	3	3
0.22	1660	1663	1667	1671	1675	1679	1683	1687	1690	1694	0	1	1	2	2	2	3	3	3
0.23	1698	1702	1706	1710	1714	1718	1722	1726	1730	1734	0	1	1	2	2	2	3	3	4
0.24	1738	1742	1746	1750	1754	1758	1762	1766	1770	1774	0	1	1	2	2	2	3	3	4
0.25	1778	1782	1786	1791	1795	1799	1803	1807	1811	1816	0	1	1	2	2	2	3	3	4
0.26	1820	1824	1828	1832	1837	1841	1845	1849	1854	1858	0	1	1	2	2	3	3	3	4
0.27	1862	1866	1871	1875	1879	1884	1888	1892	1897	1901	0	1	1	2	2	3	3	3	4
0.28	1905	1910	1914	1919	1923	1928	1932	1936	1941	1945	0	1	1	2	2	3	3	4	4
0.29	1950	1954	1959	1963	1968	1972	1977	1982	1986	1991	0	1	1	2	2	3	3	4	4
0.30	1995	2000	2004	2009	2014	2018	2023	2028	2032	2037	0	1	1	2	2	3	3	4	4
0.31	2042	2046	2051	2056	2061	2065	2070	2075	2080	2084	0	1	1	2	2	3	3	4	4
0.32	2089	2094	2099	2104	2109	2113	2118	2123	2128	2133	0	1	1	2	2	3	3	4	4
0.33	2138	2143	2148	2153	2158	2163	2168	2173	2178	2183	0	1	1	2	2	3	3	4	4
0.34	2188	2193	2198	2203	2208	2213	2218	2223	2228	2234	1	1	2	2	3	3	4	4	5
0.35	2239	2244	2249	2254	2259	2265	2270	2275	2280	2286	1	1	2	2	3	3	4	4	5
0.36	2291	2296	2301	2307	2312	2317	2323	2328	2333	2339	1	1	2	2	3	3	4	4	5
0.37	2344	2350	2355	2360	2366	2371	2377	2382	2388	2393	1	1	2	2	3	3	4	4	5
0.38	2399	2404	2410	2415	2421	2427	2432	2438	2443	2449	1	1	2	2	3	3	4	4	5
0.39	2455	2460	2466	2472	2477	2483	2489	2495	2500	2506	1	1	2	2	3	3	4	5	5
0.40	2512	2518	2523	2529	2535	2541	2547	2553	2559	2564	1	1	2	2	3	4	4	5	5
0.41	2570	2576	2582	2588	2594	2600	2606	2612	2618	2624	1	1	2	2	3	4	4	5	5
0.42	2630	2636	2642	2649	2655	2661	2667	2673	2679	2685	1	1	2	2	3	4	4	5	6
0.43	2692	2698	2704	2710	2716	2723	2729	2735	2742	2748	1	1	2	3	3	4	4	5	6
0.44	2754	2761	2767	2773	2780	2786	2793	2799	2805	2812	1	1	2	3	3	4	4	5	6
0.45	2818	2825	2831	2838	2844	2851	2858	2864	2871	2877	1	1	2	3	3	4	5	5	6
0.46	2884	2891	2897	2904	2911	2917	2924	2931	2938	2944	1	1	2	3	3	4	5	5	6
0.47	2951	2958	2965	2972	2979	2985	2992	2999	3006	3013	1	1	2	3	3	4	5	5	6
0.48	3020	3027	3034	3041	3048	3055	3062	3069	3076	3083	1	1	2	3	4	4	5	6	6
0.49	3090	3097	3105	3112	3119	3126	3133	3141	3148	3155	1	1	2	3	4	4	5	6	6

Only the mantissa (decimal part) of a logarithm is used when using the anti-log tables.

To find the number whose log is 2.2004.

Using the mantissa .2004 the number corresponding is 1586.

Since the characteristic is 2 the number must have three figures to the left of the decimal point. The number is 158.6. (Note that log 158.6 = 2.2004.)

Antilogarithms

	0	1	2	3	4	5	6	7	8	9	1	2	3	4	5	6	7	8	9
0.50	3162	3170	3177	3184	3192	3199	3206	3214	3221	3228	1	1	2	3	4	4	5	6	7
0.51	3236	3243	3251	3258	3266	3273	3281	3289	3296	3304	1	2	2	3	4	5	5	6	7
0.52	3311	3319	3327	3334	3342	3350	3357	3365	3373	3381	1	2	2	3	4	5	5	6	7
0.53	3388	3396	3404	3412	3420	3428	3436	3443	3451	3459	1	2	2	3	4	5	6	6	7
0.54	3467	3475	3483	3491	3499	3508	3516	3524	3532	3540	1	2	2	3	4	5	6	6	7
0.55	3548	3556	3565	3573	3581	3589	3597	3606	3614	3622	1	2	2	3	4	5	6	7	7
0.56	3631	3639	3648	3656	3664	3673	3681	3690	3698	3707	1	2	3	3	4	5	6	7	8
0.57	3715	3724	3733	3741	3750	3758	3767	3776	3784	3793	1	2	3	3	4	5	6	7	8
0.58	3802	3811	3819	3828	3837	3846	3855	3864	3873	3882	1	2	3	4	4	5	6	7	8
0.59	3890	3899	3908	3917	3926	3936	3945	3954	3963	3972	1	2	3	4	5	5	6	7	8
0.60	3981	3990	3999	4009	4018	4027	4036	4046	4055	4064	1	2	3	4	5	6	6	7	8
0.61	4074	4083	4093	4102	4111	4121	4130	4140	4150	4159	1	2	3	4	5	6	7	8	9
0.62	4169	4178	4188	4198	4207	4217	4227	4236	4246	4256	1	2	3	4	5	6	7	8	9
0.63	4266	4276	4285	4295	4305	4315	4325	4335	4345	4355	1	2	3	4	5	6	7	8	9
0.64	4365	4375	4385	4395	4406	4416	4426	4436	4446	4457	1	2	3	4	5	6	7	8	9
0.65	4467	4477	4487	4498	4508	4519	4529	4539	4550	4560	1	2	3	4	5	6	7	8	9
0.66	4571	4581	4592	4603	4613	4624	4634	4645	4656	4667	1	2	3	4	5	6	7	9	10
0.67	4677	4688	4699	4710	4721	4732	4742	4753	4764	4775	1	2	3	4	5	7	8	9	10
0.68	4786	4797	4808	4819	4831	4842	4853	4864	4875	4887	1	2	3	4	6	7	8	9	10
0.69	4893	4909	4920	4932	4943	4955	4966	4977	4989	5000	1	2	3	5	6	7	8	9	10
0.70	5012	5023	5035	5047	5058	5070	5082	5093	5105	5117	1	2	4	5	6	7	8	9	11
0.71	5129	5140	5152	5164	5176	5188	5200	5212	5224	5236	1	2	4	5	6	7	8	10	11
0.72	5248	5260	5272	5284	5297	5309	5321	5333	5346	5358	1	2	4	5	6	7	9	10	11
0.73	5370	5383	5395	5408	5420	5433	5445	5458	5470	5483	1	3	4	5	6	8	9	10	11
0.74	5495	5508	5521	5534	5546	5559	5572	5585	5598	5610	1	3	4	5	6	8	9	10	12
0.75	5623	5636	5649	5662	5675	5689	5702	5715	5728	5741	1	3	4	5	7	8	9	10	12
0.76	5754	5768	5781	5794	5808	5821	5834	5848	5861	5875	1	3	4	5	7	8	9	11	12
0.77	5888	5902	5916	5929	5943	5957	5970	5984	5998	6012	1	3	4	5	7	8	10	11	12
0.78	6026	6039	6053	6067	6081	6095	6109	6124	6138	6152	1	3	4	6	7	8	10	11	13
0.79	6166	6180	6194	6209	6223	6237	6252	6266	6281	6295	1	3	4	6	7	9	10	11	13
0.80	6310	6324	6339	6353	6368	6383	6397	6412	6427	6442	1	3	4	6	7	9	10	12	13
0.81	6457	6471	6486	6501	6516	6531	6546	6561	6577	6592	2	3	5	6	8	9	11	12	14
0.82	6607	6622	6637	6653	6668	6683	6699	6714	6730	6745	2	3	5	6	8	9	11	12	14
0.83	6761	6776	6792	6808	6823	6839	6855	6871	6887	6902	2	3	5	6	8	9	11	13	14
0.84	6918	6934	6950	6966	6982	6998	7015	7031	7047	7063	2	3	5	6	8	10	11	13	15
0.85	7079	7096	7112	7129	7145	7161	7178	7194	7211	7228	2	3	5	7	8	10	12	13	15
0.86	7244	7261	7278	7295	7311	7328	7345	7362	7379	7396	2	3	5	7	8	10	12	13	15
0.87	7413	7430	7447	7464	7482	7499	7516	7534	7551	7568	2	3	5	7	9	10	12	14	16
0.88	7586	7603	7621	7638	7656	7674	7691	7709	7727	7745	2	4	5	7	9	11	12	14	16
0.89	7762	7780	7798	7816	7834	7852	7870	7889	7907	7925	2	4	5	7	9	11	13	14	16
0.90	7943	7962	7980	7998	8017	8035	8054	8072	8091	8110	2	4	6	7	9	11	13	15	17
0.91	8128	8147	8166	8185	8204	8222	8241	8260	8279	8299	2	4	6	8	9	11	13	15	17
0.92	8318	8337	8356	8375	8395	8414	8433	8453	8472	8492	2	4	6	8	10	12	14	15	17
0.93	8511	8531	8551	8570	8590	8610	8630	8650	8670	8690	2	4	6	8	10	12	14	16	18
0.94	8710	8730	8750	8770	8790	8810	8831	8851	8872	8892	2	4	6	8	10	12	14	16	18
0.95	8913	8933	8954	8974	8995	9016	9036	9057	9078	9099	2	4	6	8	10	12	15	17	19
0.96	9120	9141	9162	9183	9204	9226	9247	9268	9290	9311	2	4	6	8	11	13	15	17	19
0.97	9333	9354	9376	9397	9419	9441	9462	9484	9506	9528	2	4	7	9	11	13	15	17	20
0.98	9550	9572	9594	9616	9638	9661	9683	9705	9727	9750	2	4	7	9	11	13	16	18	20
0.99	9772	9795	9817	9840	9863	9886	9908	9931	9954	9977	2	5	7	9	11	14	16	18	20

To find the number whose log is $\overline{3}.8178$.

Using the mantissa .8178 the number corresponding is 6573.

Since the characteristic is $\overline{3}$ there must be two zeros following the decimal point. The number is 0.006 573. (Note that $\log 0.006\,573$ is $\overline{3}.8178$.)

Logarithms of Sines

°	0' 0.0°	6' 0.1°	12' 0.2°	18' 0.3°	24' 0.4°	30' 0.5°	36' 0.6°	42' 0.7°	48' 0.8°	54' 0.9°	1'	2'	3'	4'	5'
0	−∞	3̄.2419	3̄.5429	3̄.7190	3̄.8439	3̄.9408	2̄.0200	2̄.0870	2̄.1450	2̄.1961	Differences				
1	2̄.2419	2̄.2832	2̄.3210	2̄.3558	2̄.3880	2̄.4179	2̄.4459	2̄.4723	2̄.4971	2̄.5206	untrustworthy				
2	2̄.5428	2̄.5640	2̄.5842	2̄.6035	2̄.6220	2̄.6397	2̄.6567	2̄.6731	2̄.6889	2̄.7041	here				
3	2̄.7188	2̄.7330	2̄.7468	2̄.7602	2̄.7731	2̄.7857	2̄.7979	2̄.8098	2̄.8213	2̄.8326					
4	2̄.8436	2̄.8543	2̄.8647	2̄.8749	2̄.8849	2̄.8946	2̄.9042	2̄.9135	2̄.9226	2̄.9315	16	32	48	64	80
5	2̄.9403	2̄.9489	2̄.9573	2̄.9655	2̄.9736	2̄.9816	2̄.9894	2̄.9970	1̄.0046	1̄.0120	13	26	39	52	65
6	1̄.0192	1̄.0264	1̄.0334	1̄.0403	1̄.0472	1̄.0539	1̄.0605	1̄.0670	1̄.0734	1̄.0797	11	22	33	44	55
7	1̄.0859	1̄.0920	1̄.0981	1̄.1040	1̄.1099	1̄.1157	1̄.1214	1̄.1271	1̄.1326	1̄.1381	10	19	29	38	48
8	1̄.1436	1̄.1489	1̄.1542	1̄.1594	1̄.1646	1̄.1697	1̄.1747	1̄.1797	1̄.1847	1̄.1895	8	17	25	34	42
9	1̄.1943	1̄.1991	1̄.2038	1̄.2085	1̄.2131	1̄.2176	1̄.2221	1̄.2266	1̄.2310	1̄.2353	8	15	23	30	38
10	1̄.2397	1̄.2439	1̄.2482	1̄.2524	1̄.2565	1̄.2606	1̄.2647	1̄.2687	1̄.2727	1̄.2767	7	14	20	27	34
11	1̄.2806	1̄.2845	1̄.2883	1̄.2921	1̄.2959	1̄.2997	1̄.3034	1̄.3070	1̄.3107	1̄.3143	6	12	19	25	31
12	1̄.3179	1̄.3214	1̄.3250	1̄.3284	1̄.3319	1̄.3353	1̄.3387	1̄.3421	1̄.3455	1̄.3488	6	11	17	23	28
13	1̄.3521	1̄.3554	1̄.3586	1̄.3618	1̄.3650	1̄.3682	1̄.3713	1̄.3745	1̄.3775	1̄.3806	5	11	16	21	26
14	1̄.3837	1̄.3867	1̄.3897	1̄.3927	1̄.3957	1̄.3986	1̄.4015	1̄.4044	1̄.4073	1̄.4102	5	10	15	20	24
15	1̄.4130	1̄.4158	1̄.4186	1̄.4214	1̄.4242	1̄.4269	1̄.4296	1̄.4323	1̄.4350	1̄.4377	5	9	14	18	23
16	1̄.4403	1̄.4430	1̄.4456	1̄.4482	1̄.4508	1̄.4533	1̄.4559	1̄.4584	1̄.4609	1̄.4634	4	9	13	17	21
17	1̄.4659	1̄.4684	1̄.4709	1̄.4733	1̄.4757	1̄.4781	1̄.4805	1̄.4829	1̄.4853	1̄.4876	4	8	12	16	20
18	1̄.4900	1̄.4923	1̄.4946	1̄.4969	1̄.4992	1̄.5015	1̄.5037	1̄.5060	1̄.5082	1̄.5104	4	8	11	15	19
19	1̄.5126	1̄.5148	1̄.5170	1̄.5192	1̄.5213	1̄.5235	1̄.5256	1̄.5278	1̄.5299	1̄.5320	4	7	11	14	18
20	1̄.5341	1̄.5361	1̄.5382	1̄.5402	1̄.5423	1̄.5443	1̄.5463	1̄.5484	1̄.5504	1̄.5523	3	7	10	13	17
21	1̄.5543	1̄.5563	1̄.5583	1̄.5602	1̄.5621	1̄.5641	1̄.5660	1̄.5679	1̄.5698	1̄.5717	3	6	10	13	16
22	1̄.5736	1̄.5754	1̄.5773	1̄.5792	1̄.5810	1̄.5828	1̄.5847	1̄.5865	1̄.5883	1̄.5901	3	6	9	12	15
23	1̄.5919	1̄.5937	1̄.5954	1̄.5972	1̄.5990	1̄.6007	1̄.6024	1̄.6042	1̄.6059	1̄.6076	3	6	9	12	14
24	1̄.6093	1̄.6110	1̄.6127	1̄.6144	1̄.6161	1̄.6177	1̄.6194	1̄.6210	1̄.6227	1̄.6243	3	6	8	11	14
25	1̄.6259	1̄.6276	1̄.6292	1̄.6308	1̄.6324	1̄.6340	1̄.6356	1̄.6371	1̄.6387	1̄.6403	3	5	8	11	13
26	1̄.6418	1̄.6434	1̄.6449	1̄.6465	1̄.6480	1̄.6495	1̄.6510	1̄.6526	1̄.6541	1̄.6556	3	5	8	10	13
27	1̄.6570	1̄.6585	1̄.6600	1̄.6615	1̄.6629	1̄.6644	1̄.6659	1̄.6673	1̄.6687	1̄.6702	2	5	7	10	12
28	1̄.6716	1̄.6730	1̄.6744	1̄.6759	1̄.6773	1̄.6787	1̄.6801	1̄.6814	1̄.6828	1̄.6842	2	5	7	9	12
29	1̄.6856	1̄.6869	1̄.6883	1̄.6896	1̄.6910	1̄.6923	1̄.6937	1̄.6950	1̄.6963	1̄.6977	2	4	7	9	11
30	1̄.6990	1̄.7003	1̄.7016	1̄.7029	1̄.7042	1̄.7055	1̄.7068	1̄.7080	1̄.7093	1̄.7106	2	4	6	9	11
31	1̄.7118	1̄.7131	1̄.7144	1̄.7156	1̄.7168	1̄.7181	1̄.7193	1̄.7205	1̄.7218	1̄.7230	2	4	6	8	10
32	1̄.7242	1̄.7254	1̄.7266	1̄.7278	1̄.7290	1̄.7302	1̄.7314	1̄.7326	1̄.7338	1̄.7349	2	4	6	8	10
33	1̄.7361	1̄.7373	1̄.7384	1̄.7396	1̄.7407	1̄.7419	1̄.7430	1̄.7442	1̄.7453	1̄.7464	2	4	6	8	10
34	1̄.7476	1̄.7487	1̄.7498	1̄.7509	1̄.7520	1̄.7531	1̄.7542	1̄.7553	1̄.7564	1̄.7575	2	4	6	7	9
35	1̄.7586	1̄.7597	1̄.7607	1̄.7618	1̄.7629	1̄.7640	1̄.7650	1̄.7661	1̄.7671	1̄.7682	2	4	5	7	9
36	1̄.7692	1̄.7703	1̄.7713	1̄.7723	1̄.7734	1̄.7744	1̄.7754	1̄.7764	1̄.7774	1̄.7785	2	3	5	7	9
37	1̄.7795	1̄.7805	1̄.7815	1̄.7825	1̄.7835	1̄.7844	1̄.7854	1̄.7864	1̄.7874	1̄.7884	2	3	5	7	8
38	1̄.7893	1̄.7903	1̄.7913	1̄.7922	1̄.7932	1̄.7941	1̄.7951	1̄.7960	1̄.7970	1̄.7979	2	3	5	6	8
39	1̄.7989	1̄.7998	1̄.8007	1̄.8017	1̄.8026	1̄.8035	1̄.8044	1̄.8053	1̄.8063	1̄.8072	2	3	5	6	8
40	1̄.8081	1̄.8090	1̄.8099	1̄.8108	1̄.8117	1̄.8125	1̄.8134	1̄.8143	1̄.8152	1̄.8161	1	3	4	6	7
41	1̄.8169	1̄.8178	1̄.8187	1̄.8195	1̄.8204	1̄.8213	1̄.8221	1̄.8230	1̄.8238	1̄.8247	1	3	4	6	7
42	1̄.8255	1̄.8264	1̄.8272	1̄.8280	1̄.8289	1̄.8297	1̄.8305	1̄.8313	1̄.8322	1̄.8330	1	3	4	6	7
43	1̄.8338	1̄.8346	1̄.8354	1̄.8362	1̄.8370	1̄.8378	1̄.8386	1̄.8394	1̄.8402	1̄.8410	1	3	4	5	7
44	1̄.8418	1̄.8426	1̄.8433	1̄.8441	1̄.8449	1̄.8457	1̄.8464	1̄.8472	1̄.8480	1̄.8487	1	3	4	5	6

To find $28.25 \times \sin 39° 17'$

$$\log(28.25 \times \sin 39° 17') = \log 28.25 + \log \sin 39° 17'$$
$$= 1.4510 + \bar{1}.8015 = 1.2525$$

From the anti-log tables the answer is 17.88.

Logarithms of Sines

°	0' 0.0°	6' 0.1°	12' 0.2°	18' 0.3°	24' 0.4°	30' 0.5°	36' 0.6°	42' 0.7°	48' 0.8°	54' 0.9°	1'	2'	3'	4'	5'
45	$\bar{1}$.8495	$\bar{1}$.8502	$\bar{1}$.8510	$\bar{1}$.8517	$\bar{1}$.8525	$\bar{1}$.8532	$\bar{1}$.8540	$\bar{1}$.8547	$\bar{1}$.8555	$\bar{1}$.8562	1	2	4	5	6
46	$\bar{1}$.8569	$\bar{1}$.8577	$\bar{1}$.8584	$\bar{1}$.8591	$\bar{1}$.8598	$\bar{1}$.8606	$\bar{1}$.8613	$\bar{1}$.8620	$\bar{1}$.8627	$\bar{1}$.8634	1	2	4	5	6
47	$\bar{1}$.8641	$\bar{1}$.8648	$\bar{1}$.8655	$\bar{1}$.8662	$\bar{1}$.8669	$\bar{1}$.8676	$\bar{1}$.8683	$\bar{1}$.8690	$\bar{1}$.8697	$\bar{1}$.8704	1	2	3	5	6
48	$\bar{1}$.8711	$\bar{1}$.8718	$\bar{1}$.8724	$\bar{1}$.8731	$\bar{1}$.8738	$\bar{1}$.8745	$\bar{1}$.8751	$\bar{1}$.8758	$\bar{1}$.8765	$\bar{1}$.8771	1	2	3	4	6
49	$\bar{1}$.8778	$\bar{1}$.8784	$\bar{1}$.8791	$\bar{1}$.8797	$\bar{1}$.8804	$\bar{1}$.8810	$\bar{1}$.8817	$\bar{1}$.8823	$\bar{1}$.8830	$\bar{1}$.8836	1	2	3	4	5
50	$\bar{1}$.8843	$\bar{1}$.8849	$\bar{1}$.8855	$\bar{1}$.8862	$\bar{1}$.8868	$\bar{1}$.8874	$\bar{1}$.8880	$\bar{1}$.8887	$\bar{1}$.8893	$\bar{1}$.8899	1	2	3	4	5
51	$\bar{1}$.8905	$\bar{1}$.8911	$\bar{1}$.8917	$\bar{1}$.8923	$\bar{1}$.8929	$\bar{1}$.8935	$\bar{1}$.8941	$\bar{1}$.8947	$\bar{1}$.8953	$\bar{1}$.8959	1	2	3	4	5
52	$\bar{1}$.8965	$\bar{1}$.8971	$\bar{1}$.8977	$\bar{1}$.8983	$\bar{1}$.8989	$\bar{1}$.8995	$\bar{1}$.9000	$\bar{1}$.9006	$\bar{1}$.9012	$\bar{1}$.9018	1	2	3	4	5
53	$\bar{1}$.9023	$\bar{1}$.9029	$\bar{1}$.9035	$\bar{1}$.9041	$\bar{1}$.9046	$\bar{1}$.9052	$\bar{1}$.9057	$\bar{1}$.9063	$\bar{1}$.9069	$\bar{1}$.9074	1	2	3	4	5
54	$\bar{1}$.9080	$\bar{1}$.9085	$\bar{1}$.9091	$\bar{1}$.9096	$\bar{1}$.9101	$\bar{1}$.9107	$\bar{1}$.9112	$\bar{1}$.9118	$\bar{1}$.9123	$\bar{1}$.9128	1	2	3	4	5
55	$\bar{1}$.9134	$\bar{1}$.9139	$\bar{1}$.9144	$\bar{1}$.9149	$\bar{1}$.9155	$\bar{1}$.9160	$\bar{1}$.9165	$\bar{1}$.9170	$\bar{1}$.9175	$\bar{1}$.9181	1	2	3	3	4
56	$\bar{1}$.9186	$\bar{1}$.9191	$\bar{1}$.9196	$\bar{1}$.9201	$\bar{1}$.9206	$\bar{1}$.9211	$\bar{1}$.9216	$\bar{1}$.9221	$\bar{1}$.9226	$\bar{1}$.9231	1	2	3	3	4
57	$\bar{1}$.9236	$\bar{1}$.9241	$\bar{1}$.9246	$\bar{1}$.9251	$\bar{1}$.9255	$\bar{1}$.9260	$\bar{1}$.9265	$\bar{1}$.9270	$\bar{1}$.9275	$\bar{1}$.9279	1	2	2	3	4
58	$\bar{1}$.9284	$\bar{1}$.9289	$\bar{1}$.9294	$\bar{1}$.9298	$\bar{1}$.9303	$\bar{1}$.9308	$\bar{1}$.9312	$\bar{1}$.9317	$\bar{1}$.9322	$\bar{1}$.9326	1	2	2	3	4
59	$\bar{1}$.9331	$\bar{1}$.9335	$\bar{1}$.9340	$\bar{1}$.9344	$\bar{1}$.9349	$\bar{1}$.9353	$\bar{1}$.9358	$\bar{1}$.9362	$\bar{1}$.9367	$\bar{1}$.9371	1	1	2	3	4
60	$\bar{1}$.9375	$\bar{1}$.9380	$\bar{1}$.9384	$\bar{1}$.9388	$\bar{1}$.9393	$\bar{1}$.9397	$\bar{1}$.9401	$\bar{1}$.9406	$\bar{1}$.9410	$\bar{1}$.9414	1	1	2	3	4
61	$\bar{1}$.9418	$\bar{1}$.9422	$\bar{1}$.9427	$\bar{1}$.9431	$\bar{1}$.9435	$\bar{1}$.9439	$\bar{1}$.9443	$\bar{1}$.9447	$\bar{1}$.9451	$\bar{1}$.9455	1	1	2	3	3
62	$\bar{1}$.9459	$\bar{1}$.9463	$\bar{1}$.9467	$\bar{1}$.9471	$\bar{1}$.9475	$\bar{1}$.9479	$\bar{1}$.9483	$\bar{1}$.9487	$\bar{1}$.9491	$\bar{1}$.9495	1	1	2	3	3
63	$\bar{1}$.9499	$\bar{1}$.9503	$\bar{1}$.9506	$\bar{1}$.9510	$\bar{1}$.9514	$\bar{1}$.9518	$\bar{1}$.9522	$\bar{1}$.9525	$\bar{1}$.9529	$\bar{1}$.9533	1	1	2	3	3
64	$\bar{1}$.9537	$\bar{1}$.9540	$\bar{1}$.9544	$\bar{1}$.9548	$\bar{1}$.9551	$\bar{1}$.9555	$\bar{1}$.9558	$\bar{1}$.9562	$\bar{1}$.9566	$\bar{1}$.9569	1	1	2	2	3
65	$\bar{1}$.9573	$\bar{1}$.9576	$\bar{1}$.9580	$\bar{1}$.9583	$\bar{1}$.9587	$\bar{1}$.9590	$\bar{1}$.9594	$\bar{1}$.9597	$\bar{1}$.9601	$\bar{1}$.9604	1	1	2	2	3
66	$\bar{1}$.9607	$\bar{1}$.9611	$\bar{1}$.9614	$\bar{1}$.9617	$\bar{1}$.9621	$\bar{1}$.9624	$\bar{1}$.9627	$\bar{1}$.9631	$\bar{1}$.9634	$\bar{1}$.9637	1	1	2	2	3
67	$\bar{1}$.9640	$\bar{1}$.9643	$\bar{1}$.9647	$\bar{1}$.9650	$\bar{1}$.9653	$\bar{1}$.9656	$\bar{1}$.9659	$\bar{1}$.9662	$\bar{1}$.9666	$\bar{1}$.9669	1	1	2	2	3
68	$\bar{1}$.9672	$\bar{1}$.9675	$\bar{1}$.9678	$\bar{1}$.9681	$\bar{1}$.9684	$\bar{1}$.9687	$\bar{1}$.9690	$\bar{1}$.9693	$\bar{1}$.9696	$\bar{1}$.9699	1	1	2	2	2
69	$\bar{1}$.9702	$\bar{1}$.9704	$\bar{1}$.9707	$\bar{1}$.9710	$\bar{1}$.9713	$\bar{1}$.9716	$\bar{1}$.9719	$\bar{1}$.9722	$\bar{1}$.9724	$\bar{1}$.9727	0	1	1	2	2
70	$\bar{1}$.9730	$\bar{1}$.9733	$\bar{1}$.9735	$\bar{1}$.9738	$\bar{1}$.9741	$\bar{1}$.9743	$\bar{1}$.9746	$\bar{1}$.9749	$\bar{1}$.9751	$\bar{1}$.9754	0	1	1	2	2
71	$\bar{1}$.9757	$\bar{1}$.9759	$\bar{1}$.9762	$\bar{1}$.9764	$\bar{1}$.9767	$\bar{1}$.9770	$\bar{1}$.9772	$\bar{1}$.9775	$\bar{1}$.9777	$\bar{1}$.9780	0	1	1	2	2
72	$\bar{1}$.9782	$\bar{1}$.9785	$\bar{1}$.9787	$\bar{1}$.9789	$\bar{1}$.9792	$\bar{1}$.9794	$\bar{1}$.9797	$\bar{1}$.9799	$\bar{1}$.9801	$\bar{1}$.9804	0	1	1	2	2
73	$\bar{1}$.9806	$\bar{1}$.9808	$\bar{1}$.9811	$\bar{1}$.9813	$\bar{1}$.9815	$\bar{1}$.9817	$\bar{1}$.9820	$\bar{1}$.9822	$\bar{1}$.9824	$\bar{1}$.9826	0	1	1	1	2
74	$\bar{1}$.9828	$\bar{1}$.9831	$\bar{1}$.9833	$\bar{1}$.9835	$\bar{1}$.9837	$\bar{1}$.9839	$\bar{1}$.9841	$\bar{1}$.9843	$\bar{1}$.9845	$\bar{1}$.9847	0	1	1	1	2
75	$\bar{1}$.9849	$\bar{1}$.9851	$\bar{1}$.9853	$\bar{1}$.9855	$\bar{1}$.9857	$\bar{1}$.9859	$\bar{1}$.9861	$\bar{1}$.9863	$\bar{1}$.9865	$\bar{1}$.9867	0	1	1	1	2
76	$\bar{1}$.9869	$\bar{1}$.9871	$\bar{1}$.9873	$\bar{1}$.9875	$\bar{1}$.9876	$\bar{1}$.9878	$\bar{1}$.9880	$\bar{1}$.9882	$\bar{1}$.9884	$\bar{1}$.9885	0	1	1	1	2
77	$\bar{1}$.9887	$\bar{1}$.9889	$\bar{1}$.9891	$\bar{1}$.9892	$\bar{1}$.9894	$\bar{1}$.9896	$\bar{1}$.9897	$\bar{1}$.9899	$\bar{1}$.9901	$\bar{1}$.9902	0	1	1	1	1
78	$\bar{1}$.9904	$\bar{1}$.9906	$\bar{1}$.9907	$\bar{1}$.9909	$\bar{1}$.9910	$\bar{1}$.9912	$\bar{1}$.9913	$\bar{1}$.9915	$\bar{1}$.9916	$\bar{1}$.9918	0	1	1	1	1
79	$\bar{1}$.9919	$\bar{1}$.9921	$\bar{1}$.9922	$\bar{1}$.9924	$\bar{1}$.9925	$\bar{1}$.9927	$\bar{1}$.9928	$\bar{1}$.9929	$\bar{1}$.9931	$\bar{1}$.9932	0	0	1	1	1
80	$\bar{1}$.9934	$\bar{1}$.9935	$\bar{1}$.9936	$\bar{1}$.9937	$\bar{1}$.9939	$\bar{1}$.9940	$\bar{1}$.9941	$\bar{1}$.9943	$\bar{1}$.9944	$\bar{1}$.9945	0	0	1	1	1
81	$\bar{1}$.9946	$\bar{1}$.9947	$\bar{1}$.9949	$\bar{1}$.9950	$\bar{1}$.9951	$\bar{1}$.9952	$\bar{1}$.9953	$\bar{1}$.9954	$\bar{1}$.9955	$\bar{1}$.9956	0	0	1	1	1
82	$\bar{1}$.9958	$\bar{1}$.9959	$\bar{1}$.9960	$\bar{1}$.9961	$\bar{1}$.9962	$\bar{1}$.9963	$\bar{1}$.9964	$\bar{1}$.9965	$\bar{1}$.9966	$\bar{1}$.9967	0	0	0	1	1
83	$\bar{1}$.9968	$\bar{1}$.9968	$\bar{1}$.9969	$\bar{1}$.9970	$\bar{1}$.9971	$\bar{1}$.9972	$\bar{1}$.9973	$\bar{1}$.9974	$\bar{1}$.9975	$\bar{1}$.9975	0	0	0	1	1
84	$\bar{1}$.9976	$\bar{1}$.9977	$\bar{1}$.9978	$\bar{1}$.9978	$\bar{1}$.9979	$\bar{1}$.9980	$\bar{1}$.9981	$\bar{1}$.9981	$\bar{1}$.9982	$\bar{1}$.9983	0	0	0	0	1
85	$\bar{1}$.9983	$\bar{1}$.9984	$\bar{1}$.9985	$\bar{1}$.9985	$\bar{1}$.9986	$\bar{1}$.9987	$\bar{1}$.9987	$\bar{1}$.9988	$\bar{1}$.9988	$\bar{1}$.9989	0	0	0	0	0
86	$\bar{1}$.9989	$\bar{1}$.9990	$\bar{1}$.9990	$\bar{1}$.9991	$\bar{1}$.9991	$\bar{1}$.9992	$\bar{1}$.9992	$\bar{1}$.9993	$\bar{1}$.9993	$\bar{1}$.9994	0	0	0	0	0
87	$\bar{1}$.9994	$\bar{1}$.9994	$\bar{1}$.9995	$\bar{1}$.9995	$\bar{1}$.9996	$\bar{1}$.9996	$\bar{1}$.9996	$\bar{1}$.9996	$\bar{1}$.9997	$\bar{1}$.9997	0	0	0	0	0
88	$\bar{1}$.9997	$\bar{1}$.9998	$\bar{1}$.9998	$\bar{1}$.9998	$\bar{1}$.9998	$\bar{1}$.9999	$\bar{1}$.9999	$\bar{1}$.9999	$\bar{1}$.9999	$\bar{1}$.9999	0	0	0	0	0
89	$\bar{1}$.9999	$\bar{1}$.9999	0.0000	0.0000	0.0000	0.0000	0.0000	0.0000	0.0000	0.0000					
90	0.0000														

To find the angle A given that $\sin A = \dfrac{19.16}{23.45}$

$$\log \sin A = \log\left(\frac{19.16}{23.45}\right) = \log 19.16 - \log 23.45$$
$$= 1.2824 - 1.3701 = \bar{1}.9123$$

using the log sine tables

$$A = 54° \ 48'$$

25

Logarithms of Cosines

Numbers in difference columns to be *subtracted*, not added.

°	0' 0.0°	6' 0.1°	12' 0.2°	18' 0.3°	24' 0.4°	30' 0.5°	36' 0.6°	42' 0.7°	48' 0.8°	54' 0.9°	1'	2'	3'	4'	5'
0	0.0000	0.0000	0.0000	0.0000	0.0000	0.0000	0.0000	0.0000	0.0000	$\bar{1}$.9999	0	0	0	0	0
1	$\bar{1}$.9999	$\bar{1}$.9999	$\bar{1}$.9999	$\bar{1}$.9999	$\bar{1}$.9999	$\bar{1}$.9999	$\bar{1}$.9998	$\bar{1}$.9998	$\bar{1}$.9998	$\bar{1}$.9998	0	0	0	0	0
2	$\bar{1}$.9997	$\bar{1}$.9997	$\bar{1}$.9997	$\bar{1}$.9996	$\bar{1}$.9996	$\bar{1}$.9996	$\bar{1}$.9996	$\bar{1}$.9995	$\bar{1}$.9995	$\bar{1}$.9994	0	0	0	0	0
3	$\bar{1}$.9994	$\bar{1}$.9994	$\bar{1}$.9993	$\bar{1}$.9993	$\bar{1}$.9992	$\bar{1}$.9992	$\bar{1}$.9991	$\bar{1}$.9991	$\bar{1}$.9990	$\bar{1}$.9990	0	0	0	0	0
4	$\bar{1}$.9989	$\bar{1}$.9989	$\bar{1}$.9988	$\bar{1}$.9988	$\bar{1}$.9987	$\bar{1}$.9987	$\bar{1}$.9986	$\bar{1}$.9985	$\bar{1}$.9985	$\bar{1}$.9984	0	0	0	0	0
5	$\bar{1}$.9983	$\bar{1}$.9983	$\bar{1}$.9982	$\bar{1}$.9981	$\bar{1}$.9981	$\bar{1}$.9980	$\bar{1}$.9979	$\bar{1}$.9978	$\bar{1}$.9978	$\bar{1}$.9977	0	0	0	0	1
6	$\bar{1}$.9976	$\bar{1}$.9975	$\bar{1}$.9975	$\bar{1}$.9974	$\bar{1}$.9973	$\bar{1}$.9972	$\bar{1}$.9971	$\bar{1}$.9970	$\bar{1}$.9969	$\bar{1}$.9968	0	0	0	1	1
7	$\bar{1}$.9968	$\bar{1}$.9967	$\bar{1}$.9966	$\bar{1}$.9965	$\bar{1}$.9964	$\bar{1}$.9963	$\bar{1}$.9962	$\bar{1}$.9961	$\bar{1}$.9960	$\bar{1}$.9959	0	0	0	1	1
8	$\bar{1}$.9958	$\bar{1}$.9956	$\bar{1}$.9955	$\bar{1}$.9954	$\bar{1}$.9953	$\bar{1}$.9952	$\bar{1}$.9951	$\bar{1}$.9950	$\bar{1}$.9949	$\bar{1}$.9947	0	0	1	1	1
9	$\bar{1}$.9946	$\bar{1}$.9945	$\bar{1}$.9944	$\bar{1}$.9943	$\bar{1}$.9941	$\bar{1}$.9940	$\bar{1}$.9939	$\bar{1}$.9937	$\bar{1}$.9936	$\bar{1}$.9935	0	0	1	1	1
10	$\bar{1}$.9934	$\bar{1}$.9932	$\bar{1}$.9931	$\bar{1}$.9929	$\bar{1}$.9928	$\bar{1}$.9927	$\bar{1}$.9925	$\bar{1}$.9924	$\bar{1}$.9922	$\bar{1}$.9921	0	0	1	1	1
11	$\bar{1}$.9919	$\bar{1}$.9918	$\bar{1}$.9916	$\bar{1}$.9915	$\bar{1}$.9913	$\bar{1}$.9912	$\bar{1}$.9910	$\bar{1}$.9909	$\bar{1}$.9907	$\bar{1}$.9906	0	1	1	1	1
12	$\bar{1}$.9904	$\bar{1}$.9902	$\bar{1}$.9901	$\bar{1}$.9899	$\bar{1}$.9897	$\bar{1}$.9896	$\bar{1}$.9894	$\bar{1}$.9892	$\bar{1}$.9891	$\bar{1}$.9889	0	1	1	1	1
13	$\bar{1}$.9887	$\bar{1}$.9885	$\bar{1}$.9884	$\bar{1}$.9882	$\bar{1}$.9880	$\bar{1}$.9878	$\bar{1}$.9876	$\bar{1}$.9875	$\bar{1}$.9873	$\bar{1}$.9871	0	1	1	1	2
14	$\bar{1}$.9869	$\bar{1}$.9867	$\bar{1}$.9865	$\bar{1}$.9863	$\bar{1}$.9861	$\bar{1}$.9859	$\bar{1}$.9857	$\bar{1}$.9855	$\bar{1}$.9853	$\bar{1}$.9851	0	1	1	1	2
15	$\bar{1}$.9849	$\bar{1}$.9847	$\bar{1}$.9845	$\bar{1}$.9843	$\bar{1}$.9841	$\bar{1}$.9839	$\bar{1}$.9837	$\bar{1}$.9835	$\bar{1}$.9833	$\bar{1}$.9831	0	1	1	1	2
16	$\bar{1}$.9828	$\bar{1}$.9826	$\bar{1}$.9824	$\bar{1}$.9822	$\bar{1}$.9820	$\bar{1}$.9817	$\bar{1}$.9815	$\bar{1}$.9813	$\bar{1}$.9811	$\bar{1}$.9808	0	1	1	2	2
17	$\bar{1}$.9806	$\bar{1}$.9804	$\bar{1}$.9801	$\bar{1}$.9799	$\bar{1}$.9797	$\bar{1}$.9794	$\bar{1}$.9792	$\bar{1}$.9789	$\bar{1}$.9787	$\bar{1}$.9785	0	1	1	2	2
18	$\bar{1}$.9782	$\bar{1}$.9780	$\bar{1}$.9777	$\bar{1}$.9775	$\bar{1}$.9772	$\bar{1}$.9770	$\bar{1}$.9767	$\bar{1}$.9764	$\bar{1}$.9762	$\bar{1}$.9759	0	1	1	2	2
19	$\bar{1}$.9757	$\bar{1}$.9754	$\bar{1}$.9751	$\bar{1}$.9749	$\bar{1}$.9746	$\bar{1}$.9743	$\bar{1}$.9741	$\bar{1}$.9738	$\bar{1}$.9735	$\bar{1}$.9733	0	1	1	2	2
20	$\bar{1}$.9730	$\bar{1}$.9727	$\bar{1}$.9724	$\bar{1}$.9722	$\bar{1}$.9719	$\bar{1}$.9716	$\bar{1}$.9713	$\bar{1}$.9710	$\bar{1}$.9707	$\bar{1}$.9704	0	1	1	2	2
21	$\bar{1}$.9702	$\bar{1}$.9699	$\bar{1}$.9696	$\bar{1}$.9693	$\bar{1}$.9690	$\bar{1}$.9687	$\bar{1}$.9684	$\bar{1}$.9681	$\bar{1}$.9678	$\bar{1}$.9675	1	1	2	2	2
22	$\bar{1}$.9672	$\bar{1}$.9669	$\bar{1}$.9666	$\bar{1}$.9662	$\bar{1}$.9659	$\bar{1}$.9656	$\bar{1}$.9653	$\bar{1}$.9650	$\bar{1}$.9647	$\bar{1}$.9643	1	1	2	2	3
23	$\bar{1}$.9640	$\bar{1}$.9637	$\bar{1}$.9634	$\bar{1}$.9631	$\bar{1}$.9627	$\bar{1}$.9624	$\bar{1}$.9621	$\bar{1}$.9617	$\bar{1}$.9614	$\bar{1}$.9611	1	1	2	2	3
24	$\bar{1}$.9607	$\bar{1}$.9604	$\bar{1}$.9601	$\bar{1}$.9597	$\bar{1}$.9594	$\bar{1}$.9590	$\bar{1}$.9587	$\bar{1}$.9583	$\bar{1}$.9580	$\bar{1}$.9576	1	1	2	2	3
25	$\bar{1}$.9573	$\bar{1}$.9569	$\bar{1}$.9566	$\bar{1}$.9562	$\bar{1}$.9558	$\bar{1}$.9555	$\bar{1}$.9551	$\bar{1}$.9548	$\bar{1}$.9544	$\bar{1}$.9540	1	1	2	2	3
26	$\bar{1}$.9537	$\bar{1}$.9533	$\bar{1}$.9529	$\bar{1}$.9525	$\bar{1}$.9522	$\bar{1}$.9518	$\bar{1}$.9514	$\bar{1}$.9510	$\bar{1}$.9506	$\bar{1}$.9503	1	1	2	3	3
27	$\bar{1}$.9499	$\bar{1}$.9495	$\bar{1}$.9491	$\bar{1}$.9487	$\bar{1}$.9483	$\bar{1}$.9479	$\bar{1}$.9475	$\bar{1}$.9471	$\bar{1}$.9467	$\bar{1}$.9463	1	1	2	3	3
28	$\bar{1}$.9459	$\bar{1}$.9455	$\bar{1}$.9451	$\bar{1}$.9447	$\bar{1}$.9443	$\bar{1}$.9439	$\bar{1}$.9435	$\bar{1}$.9431	$\bar{1}$.9427	$\bar{1}$.9422	1	1	2	3	3
29	$\bar{1}$.9418	$\bar{1}$.9414	$\bar{1}$.9410	$\bar{1}$.9406	$\bar{1}$.9401	$\bar{1}$.9397	$\bar{1}$.9393	$\bar{1}$.9388	$\bar{1}$.9384	$\bar{1}$.9380	1	1	2	3	4
30	$\bar{1}$.9375	$\bar{1}$.9371	$\bar{1}$.9367	$\bar{1}$.9362	$\bar{1}$.9358	$\bar{1}$.9353	$\bar{1}$.9349	$\bar{1}$.9344	$\bar{1}$.9340	$\bar{1}$.9335	1	1	2	3	4
31	$\bar{1}$.9331	$\bar{1}$.9326	$\bar{1}$.9322	$\bar{1}$.9317	$\bar{1}$.9312	$\bar{1}$.9308	$\bar{1}$.9303	$\bar{1}$.9298	$\bar{1}$.9294	$\bar{1}$.9289	1	2	2	3	4
32	$\bar{1}$.9284	$\bar{1}$.9279	$\bar{1}$.9275	$\bar{1}$.9270	$\bar{1}$.9265	$\bar{1}$.9260	$\bar{1}$.9255	$\bar{1}$.9251	$\bar{1}$.9246	$\bar{1}$.9241	1	2	2	3	4
33	$\bar{1}$.9236	$\bar{1}$.9231	$\bar{1}$.9226	$\bar{1}$.9221	$\bar{1}$.9216	$\bar{1}$.9211	$\bar{1}$.9206	$\bar{1}$.9201	$\bar{1}$.9196	$\bar{1}$.9191	1	2	3	3	4
34	$\bar{1}$.9186	$\bar{1}$.9181	$\bar{1}$.9175	$\bar{1}$.9170	$\bar{1}$.9165	$\bar{1}$.9160	$\bar{1}$.9155	$\bar{1}$.9149	$\bar{1}$.9144	$\bar{1}$.9139	1	2	3	3	4
35	$\bar{1}$.9134	$\bar{1}$.9128	$\bar{1}$.9123	$\bar{1}$.9118	$\bar{1}$.9112	$\bar{1}$.9107	$\bar{1}$.9101	$\bar{1}$.9096	$\bar{1}$.9091	$\bar{1}$.9085	1	2	3	4	5
36	$\bar{1}$.9080	$\bar{1}$.9074	$\bar{1}$.9069	$\bar{1}$.9063	$\bar{1}$.9057	$\bar{1}$.9052	$\bar{1}$.9046	$\bar{1}$.9041	$\bar{1}$.9035	$\bar{1}$.9029	1	2	3	4	5
37	$\bar{1}$.9023	$\bar{1}$.9018	$\bar{1}$.9012	$\bar{1}$.9006	$\bar{1}$.9000	$\bar{1}$.8995	$\bar{1}$.8989	$\bar{1}$.8983	$\bar{1}$.8977	$\bar{1}$.8971	1	2	3	4	5
38	$\bar{1}$.8965	$\bar{1}$.8959	$\bar{1}$.8953	$\bar{1}$.8947	$\bar{1}$.8941	$\bar{1}$.8935	$\bar{1}$.8929	$\bar{1}$.8923	$\bar{1}$.8917	$\bar{1}$.8911	1	2	3	4	5
39	$\bar{1}$.8905	$\bar{1}$.8899	$\bar{1}$.8893	$\bar{1}$.8887	$\bar{1}$.8880	$\bar{1}$.8874	$\bar{1}$.8868	$\bar{1}$.8862	$\bar{1}$.8855	$\bar{1}$.8849	1	2	3	4	5
40	$\bar{1}$.8843	$\bar{1}$.8836	$\bar{1}$.8830	$\bar{1}$.8823	$\bar{1}$.8817	$\bar{1}$.8810	$\bar{1}$.8804	$\bar{1}$.8797	$\bar{1}$.8791	$\bar{1}$.8784	1	2	3	4	5
41	$\bar{1}$.8778	$\bar{1}$.8771	$\bar{1}$.8765	$\bar{1}$.8758	$\bar{1}$.8751	$\bar{1}$.8745	$\bar{1}$.8738	$\bar{1}$.8731	$\bar{1}$.8724	$\bar{1}$.8718	1	2	3	4	6
42	$\bar{1}$.8711	$\bar{1}$.8704	$\bar{1}$.8697	$\bar{1}$.8690	$\bar{1}$.8683	$\bar{1}$.8676	$\bar{1}$.8669	$\bar{1}$.8662	$\bar{1}$.8655	$\bar{1}$.8648	1	2	4	5	6
43	$\bar{1}$.8641	$\bar{1}$.8634	$\bar{1}$.8627	$\bar{1}$.8620	$\bar{1}$.8613	$\bar{1}$.8606	$\bar{1}$.8598	$\bar{1}$.8591	$\bar{1}$.8584	$\bar{1}$.8577	1	2	4	5	6
44	$\bar{1}$.8569	$\bar{1}$.8562	$\bar{1}$.8555	$\bar{1}$.8547	$\bar{1}$.8540	$\bar{1}$.8532	$\bar{1}$.8525	$\bar{1}$.8517	$\bar{1}$.8510	$\bar{1}$.8502	1	2	4	5	6

To find $23.05 \times \cos 16° 51'$.

$$\log(23.05 \times \cos 16° 51') = \log 23.05 + \log \cos 16° 51'$$
$$= 1.3626 + \bar{1}.9810$$
$$= 1.3436$$

From the anti-log tables the answer is 22.06.

Logarithms of Cosines

Numbers in difference columns to be *subtracted*, not added.

°	0' 0.0°	6' 0.1°	12' 0.2°	18' 0.3°	24' 0.4°	30' 0.5°	36' 0.6°	42' 0.7°	48' 0.8°	54' 0.9°	1'	2'	3'	4'	5'
45	$\bar{1}$.8495	$\bar{1}$.8487	$\bar{1}$.8480	$\bar{1}$.8472	$\bar{1}$.8464	$\bar{1}$.8457	$\bar{1}$.8449	$\bar{1}$.8441	$\bar{1}$.8433	$\bar{1}$.8426	1	3	4	5	6
46	$\bar{1}$.8418	$\bar{1}$.8410	$\bar{1}$.8402	$\bar{1}$.8394	$\bar{1}$.8386	$\bar{1}$.8378	$\bar{1}$.8370	$\bar{1}$.8362	$\bar{1}$.8354	$\bar{1}$.8346	1	3	4	5	7
47	$\bar{1}$.8338	$\bar{1}$.8330	$\bar{1}$.8322	$\bar{1}$.8313	$\bar{1}$.8305	$\bar{1}$.8297	$\bar{1}$.8289	$\bar{1}$.8280	$\bar{1}$.8272	$\bar{1}$.8264	1	3	4	6	7
48	$\bar{1}$.8255	$\bar{1}$.8247	$\bar{1}$.8238	$\bar{1}$.8230	$\bar{1}$.8221	$\bar{1}$.8213	$\bar{1}$.8204	$\bar{1}$.8195	$\bar{1}$.8187	$\bar{1}$.8178	1	3	4	6	7
49	$\bar{1}$.8169	$\bar{1}$.8161	$\bar{1}$.8152	$\bar{1}$.8143	$\bar{1}$.8134	$\bar{1}$.8125	$\bar{1}$.8117	$\bar{1}$.8108	$\bar{1}$.8099	$\bar{1}$.8090	1	3	4	6	7
50	$\bar{1}$.8081	$\bar{1}$.8072	$\bar{1}$.8063	$\bar{1}$.8053	$\bar{1}$.8044	$\bar{1}$.8035	$\bar{1}$.8026	$\bar{1}$.8017	$\bar{1}$.8007	$\bar{1}$.7998	2	3	5	6	8
51	$\bar{1}$.7989	$\bar{1}$.7979	$\bar{1}$.7970	$\bar{1}$.7960	$\bar{1}$.7951	$\bar{1}$.7941	$\bar{1}$.7932	$\bar{1}$.7922	$\bar{1}$.7913	$\bar{1}$.7903	2	3	5	6	8
52	$\bar{1}$.7893	$\bar{1}$.7884	$\bar{1}$.7874	$\bar{1}$.7864	$\bar{1}$.7854	$\bar{1}$.7844	$\bar{1}$.7835	$\bar{1}$.7825	$\bar{1}$.7815	$\bar{1}$.7805	2	3	5	7	8
53	$\bar{1}$.7795	$\bar{1}$.7785	$\bar{1}$.7774	$\bar{1}$.7764	$\bar{1}$.7754	$\bar{1}$.7744	$\bar{1}$.7734	$\bar{1}$.7723	$\bar{1}$.7713	$\bar{1}$.7703	2	3	5	7	9
54	$\bar{1}$.7692	$\bar{1}$.7682	$\bar{1}$.7671	$\bar{1}$.7661	$\bar{1}$.7650	$\bar{1}$.7640	$\bar{1}$.7629	$\bar{1}$.7618	$\bar{1}$.7607	$\bar{1}$.7597	2	4	5	7	9
55	$\bar{1}$.7586	$\bar{1}$.7575	$\bar{1}$.7564	$\bar{1}$.7553	$\bar{1}$.7542	$\bar{1}$.7531	$\bar{1}$.7520	$\bar{1}$.7509	$\bar{1}$.7498	$\bar{1}$.7487	2	4	6	7	9
56	$\bar{1}$.7476	$\bar{1}$.7464	$\bar{1}$.7453	$\bar{1}$.7442	$\bar{1}$.7430	$\bar{1}$.7419	$\bar{1}$.7407	$\bar{1}$.7396	$\bar{1}$.7384	$\bar{1}$.7373	2	4	6	8	10
57	$\bar{1}$.7361	$\bar{1}$.7349	$\bar{1}$.7338	$\bar{1}$.7326	$\bar{1}$.7314	$\bar{1}$.7302	$\bar{1}$.7290	$\bar{1}$.7278	$\bar{1}$.7266	$\bar{1}$.7254	2	4	6	8	10
58	$\bar{1}$.7242	$\bar{1}$.7230	$\bar{1}$.7218	$\bar{1}$.7205	$\bar{1}$.7193	$\bar{1}$.7181	$\bar{1}$.7168	$\bar{1}$.7156	$\bar{1}$.7144	$\bar{1}$.7131	2	4	6	8	10
59	$\bar{1}$.7118	$\bar{1}$.7106	$\bar{1}$.7093	$\bar{1}$.7080	$\bar{1}$.7068	$\bar{1}$.7055	$\bar{1}$.7042	$\bar{1}$.7029	$\bar{1}$.7016	$\bar{1}$.7003	2	4	6	9	11
60	$\bar{1}$.6990	$\bar{1}$.6977	$\bar{1}$.6963	$\bar{1}$.6950	$\bar{1}$.6937	$\bar{1}$.6923	$\bar{1}$.6910	$\bar{1}$.6896	$\bar{1}$.6883	$\bar{1}$.6869	2	4	7	9	11
61	$\bar{1}$.6856	$\bar{1}$.6842	$\bar{1}$.6828	$\bar{1}$.6814	$\bar{1}$.6801	$\bar{1}$.6787	$\bar{1}$.6773	$\bar{1}$.6759	$\bar{1}$.6744	$\bar{1}$.6730	2	5	7	9	12
62	$\bar{1}$.6716	$\bar{1}$.6702	$\bar{1}$.6687	$\bar{1}$.6673	$\bar{1}$.6659	$\bar{1}$.6644	$\bar{1}$.6629	$\bar{1}$.6615	$\bar{1}$.6600	$\bar{1}$.6585	2	5	7	10	12
63	$\bar{1}$.6570	$\bar{1}$.6556	$\bar{1}$.6541	$\bar{1}$.6526	$\bar{1}$.6510	$\bar{1}$.6495	$\bar{1}$.6480	$\bar{1}$.6465	$\bar{1}$.6449	$\bar{1}$.6434	3	5	8	10	13
64	$\bar{1}$.6418	$\bar{1}$.6403	$\bar{1}$.6387	$\bar{1}$.6371	$\bar{1}$.6356	$\bar{1}$.6340	$\bar{1}$.6324	$\bar{1}$.6308	$\bar{1}$.6292	$\bar{1}$.6276	3	5	8	11	13
65	$\bar{1}$.6259	$\bar{1}$.6243	$\bar{1}$.6227	$\bar{1}$.6210	$\bar{1}$.6194	$\bar{1}$.6177	$\bar{1}$.6161	$\bar{1}$.6144	$\bar{1}$.6127	$\bar{1}$.6110	3	6	8	11	14
66	$\bar{1}$.6093	$\bar{1}$.6076	$\bar{1}$.6059	$\bar{1}$.6042	$\bar{1}$.6024	$\bar{1}$.6007	$\bar{1}$.5990	$\bar{1}$.5972	$\bar{1}$.5954	$\bar{1}$.5937	3	6	9	12	14
67	$\bar{1}$.5919	$\bar{1}$.5901	$\bar{1}$.5883	$\bar{1}$.5865	$\bar{1}$.5847	$\bar{1}$.5828	$\bar{1}$.5810	$\bar{1}$.5792	$\bar{1}$.5773	$\bar{1}$.5754	3	6	9	12	15
68	$\bar{1}$.5736	$\bar{1}$.5717	$\bar{1}$.5698	$\bar{1}$.5679	$\bar{1}$.5660	$\bar{1}$.5641	$\bar{1}$.5621	$\bar{1}$.5602	$\bar{1}$.5583	$\bar{1}$.5563	3	6	10	13	16
69	$\bar{1}$.5543	$\bar{1}$.5523	$\bar{1}$.5504	$\bar{1}$.5484	$\bar{1}$.5463	$\bar{1}$.5443	$\bar{1}$.5423	$\bar{1}$.5402	$\bar{1}$.5382	$\bar{1}$.5361	3	7	10	13	17
70	$\bar{1}$.5341	$\bar{1}$.5320	$\bar{1}$.5299	$\bar{1}$.5278	$\bar{1}$.5256	$\bar{1}$.5235	$\bar{1}$.5213	$\bar{1}$.5192	$\bar{1}$.5170	$\bar{1}$.5148	4	7	11	14	18
71	$\bar{1}$.5126	$\bar{1}$.5104	$\bar{1}$.5082	$\bar{1}$.5060	$\bar{1}$.5037	$\bar{1}$.5015	$\bar{1}$.4992	$\bar{1}$.4969	$\bar{1}$.4946	$\bar{1}$.4923	4	8	11	15	19
72	$\bar{1}$.4900	$\bar{1}$.4876	$\bar{1}$.4853	$\bar{1}$.4829	$\bar{1}$.4805	$\bar{1}$.4781	$\bar{1}$.4757	$\bar{1}$.4733	$\bar{1}$.4709	$\bar{1}$.4684	4	8	12	16	20
73	$\bar{1}$.4659	$\bar{1}$.4634	$\bar{1}$.4609	$\bar{1}$.4584	$\bar{1}$.4559	$\bar{1}$.4533	$\bar{1}$.4508	$\bar{1}$.4482	$\bar{1}$.4456	$\bar{1}$.4430	4	9	13	17	21
74	$\bar{1}$.4403	$\bar{1}$.4377	$\bar{1}$.4350	$\bar{1}$.4323	$\bar{1}$.4296	$\bar{1}$.4269	$\bar{1}$.4242	$\bar{1}$.4214	$\bar{1}$.4186	$\bar{1}$.4158	5	9	14	18	23
75	$\bar{1}$.4130	$\bar{1}$.4102	$\bar{1}$.4073	$\bar{1}$.4044	$\bar{1}$.4015	$\bar{1}$.3986	$\bar{1}$.3957	$\bar{1}$.3927	$\bar{1}$.3897	$\bar{1}$.3867	5	10	15	20	24
76	$\bar{1}$.3837	$\bar{1}$.3806	$\bar{1}$.3775	$\bar{1}$.3745	$\bar{1}$.3713	$\bar{1}$.3682	$\bar{1}$.3650	$\bar{1}$.3618	$\bar{1}$.3586	$\bar{1}$.3554	5	11	16	21	26
77	$\bar{1}$.3521	$\bar{1}$.3488	$\bar{1}$.3455	$\bar{1}$.3421	$\bar{1}$.3387	$\bar{1}$.3353	$\bar{1}$.3319	$\bar{1}$.3284	$\bar{1}$.3250	$\bar{1}$.3214	6	11	17	23	28
78	$\bar{1}$.3179	$\bar{1}$.3143	$\bar{1}$.3107	$\bar{1}$.3070	$\bar{1}$.3034	$\bar{1}$.2997	$\bar{1}$.2959	$\bar{1}$.2921	$\bar{1}$.2883	$\bar{1}$.2845	6	12	19	25	31
79	$\bar{1}$.2806	$\bar{1}$.2767	$\bar{1}$.2727	$\bar{1}$.2687	$\bar{1}$.2647	$\bar{1}$.2606	$\bar{1}$.2565	$\bar{1}$.2524	$\bar{1}$.2482	$\bar{1}$.2439	7	14	20	27	34
80	$\bar{1}$.2397	$\bar{1}$.2353	$\bar{1}$.2310	$\bar{1}$.2266	$\bar{1}$.2221	$\bar{1}$.2176	$\bar{1}$.2131	$\bar{1}$.2085	$\bar{1}$.2038	$\bar{1}$.1991	8	15	23	30	38
81	$\bar{1}$.1943	$\bar{1}$.1895	$\bar{1}$.1847	$\bar{1}$.1797	$\bar{1}$.1747	$\bar{1}$.1697	$\bar{1}$.1646	$\bar{1}$.1594	$\bar{1}$.1542	$\bar{1}$.1489	8	17	25	34	42
82	$\bar{1}$.1436	$\bar{1}$.1381	$\bar{1}$.1326	$\bar{1}$.1271	$\bar{1}$.1214	$\bar{1}$.1157	$\bar{1}$.1099	$\bar{1}$.1040	$\bar{1}$.0981	$\bar{1}$.0920	10	19	29	38	48
83	$\bar{1}$.0859	$\bar{1}$.0797	$\bar{1}$.0734	$\bar{1}$.0670	$\bar{1}$.0605	$\bar{1}$.0539	$\bar{1}$.0472	$\bar{1}$.0403	$\bar{1}$.0334	$\bar{1}$.0264	11	22	33	44	55
84	$\bar{1}$.0192	$\bar{1}$.0120	$\bar{1}$.0046	$\bar{2}$.9970	$\bar{2}$.9894	$\bar{2}$.9816	$\bar{2}$.9736	$\bar{2}$.9655	$\bar{2}$.9573	$\bar{2}$.9489	13	26	39	52	65
85	$\bar{2}$.9403	$\bar{2}$.9315	$\bar{2}$.9226	$\bar{2}$.9135	$\bar{2}$.9042	$\bar{2}$.8946	$\bar{2}$.8849	$\bar{2}$.8749	$\bar{2}$.8647	$\bar{2}$.8543	16	32	48	64	80
86	$\bar{2}$.8436	$\bar{2}$.8326	$\bar{2}$.8213	$\bar{2}$.8098	$\bar{2}$.7979	$\bar{2}$.7857	$\bar{2}$.7731	$\bar{2}$.7602	$\bar{2}$.7468	$\bar{2}$.7330					
87	$\bar{2}$.7188	$\bar{2}$.7041	$\bar{2}$.6889	$\bar{2}$.6731	$\bar{2}$.6567	$\bar{2}$.6397	$\bar{2}$.6220	$\bar{2}$.6035	$\bar{2}$.5842	$\bar{2}$.5640		Differences			
88	$\bar{2}$.5428	$\bar{2}$.5206	$\bar{2}$.4971	$\bar{2}$.4723	$\bar{2}$.4459	$\bar{2}$.4179	$\bar{2}$.3880	$\bar{2}$.3558	$\bar{2}$.3210	$\bar{2}$.2832		untrustworthy			
89	$\bar{2}$.2419	$\bar{2}$.1961	$\bar{2}$.1450	$\bar{2}$.0870	$\bar{2}$.0200	$\bar{3}$.9408	$\bar{3}$.8439	$\bar{3}$.7190	$\bar{3}$.5429	$\bar{3}$.2419		here			
90	$-\infty$														

To find the angle A given that $\cos A = \dfrac{20.23}{29.86}$

$$\log \cos A = \log\left(\frac{20.23}{29.86}\right) = \log 20.23 - \log 29.86$$
$$= 1.3060 - 1.4751 = \bar{1}.8309$$

Using the log cosine tables

$$A = 47° 21'$$

Logarithms of Tangents

°	0' 0.0°	6' 0.1°	12' 0.2°	18' 0.3°	24' 0.4°	30' 0.5°	36' 0.6°	42' 0.7°	48' 0.8°	54' 0.9°	1'	2'	3'	4'	5'
0	$-\infty$	$\overline{3}$.2419	$\overline{3}$.5429	$\overline{3}$.7190	$\overline{3}$.8439	$\overline{3}$.9409	$\overline{2}$.0200	$\overline{2}$.0870	$\overline{2}$.1450	$\overline{2}$.1962	\multicolumn{5}{l}{Differences}
1	$\overline{2}$.2419	$\overline{2}$.2833	$\overline{2}$.3211	$\overline{2}$.3559	$\overline{2}$.3881	$\overline{2}$.4181	$\overline{2}$.4461	$\overline{2}$.4725	$\overline{2}$.4973	$\overline{2}$.5208		untrustworthy			
2	$\overline{2}$.5431	$\overline{2}$.5643	$\overline{2}$.5845	$\overline{2}$.6038	$\overline{2}$.6223	$\overline{2}$.6401	$\overline{2}$.6571	$\overline{2}$.6736	$\overline{2}$.6894	$\overline{2}$.7046		here			
3	$\overline{2}$.7194	$\overline{2}$.7337	$\overline{2}$.7475	$\overline{2}$.7609	$\overline{2}$.7739	$\overline{2}$.7865	$\overline{2}$.7988	$\overline{2}$.8107	$\overline{2}$.8223	$\overline{2}$.8336					
4	$\overline{2}$.8446	$\overline{2}$.8554	$\overline{2}$.8659	$\overline{2}$.8762	$\overline{2}$.8862	$\overline{2}$.8960	$\overline{2}$.9056	$\overline{2}$.9150	$\overline{2}$.9241	$\overline{2}$.9331	16	32	48	64	81
5	$\overline{2}$.9420	$\overline{2}$.9506	$\overline{2}$.9591	$\overline{2}$.9674	$\overline{2}$.9756	$\overline{2}$.9836	$\overline{2}$.9915	$\overline{2}$.9992	$\overline{1}$.0068	$\overline{1}$.0143	13	26	40	53	66
6	$\overline{1}$.0216	$\overline{1}$.0289	$\overline{1}$.0360	$\overline{1}$.0430	$\overline{1}$.0499	$\overline{1}$.0567	$\overline{1}$.0633	$\overline{1}$.0699	$\overline{1}$.0764	$\overline{1}$.0828	11	22	34	45	56
7	$\overline{1}$.0891	$\overline{1}$.0954	$\overline{1}$.1015	$\overline{1}$.1076	$\overline{1}$.1135	$\overline{1}$.1194	$\overline{1}$.1252	$\overline{1}$.1310	$\overline{1}$.1367	$\overline{1}$.1423	10	20	29	39	49
8	$\overline{1}$.1478	$\overline{1}$.1533	$\overline{1}$.1587	$\overline{1}$.1640	$\overline{1}$.1693	$\overline{1}$.1745	$\overline{1}$.1797	$\overline{1}$.1848	$\overline{1}$.1898	$\overline{1}$.1948	9	17	26	35	43
9	$\overline{1}$.1997	$\overline{1}$.2046	$\overline{1}$.2094	$\overline{1}$.2142	$\overline{1}$.2189	$\overline{1}$.2236	$\overline{1}$.2282	$\overline{1}$.2328	$\overline{1}$.2374	$\overline{1}$.2419	8	16	23	31	39
10	$\overline{1}$.2463	$\overline{1}$.2507	$\overline{1}$.2551	$\overline{1}$.2594	$\overline{1}$.2637	$\overline{1}$.2680	$\overline{1}$.2722	$\overline{1}$.2764	$\overline{1}$.2805	$\overline{1}$.2846	7	14	21	28	35
11	$\overline{1}$.2887	$\overline{1}$.2927	$\overline{1}$.2967	$\overline{1}$.3006	$\overline{1}$.3046	$\overline{1}$.3085	$\overline{1}$.3123	$\overline{1}$.3162	$\overline{1}$.3200	$\overline{1}$.3237	6	13	19	26	32
12	$\overline{1}$.3275	$\overline{1}$.3312	$\overline{1}$.3349	$\overline{1}$.3385	$\overline{1}$.3422	$\overline{1}$.3458	$\overline{1}$.3493	$\overline{1}$.3529	$\overline{1}$.3564	$\overline{1}$.3599	6	12	18	24	30
13	$\overline{1}$.3634	$\overline{1}$.3668	$\overline{1}$.3702	$\overline{1}$.3736	$\overline{1}$.3770	$\overline{1}$.3804	$\overline{1}$.3837	$\overline{1}$.3870	$\overline{1}$.3903	$\overline{1}$.3935	6	11	17	22	28
14	$\overline{1}$.3968	$\overline{1}$.4000	$\overline{1}$.4032	$\overline{1}$.4064	$\overline{1}$.4095	$\overline{1}$.4127	$\overline{1}$.4158	$\overline{1}$.4189	$\overline{1}$.4220	$\overline{1}$.4250	5	10	16	21	26
15	$\overline{1}$.4281	$\overline{1}$.4311	$\overline{1}$.4341	$\overline{1}$.4371	$\overline{1}$.4400	$\overline{1}$.4430	$\overline{1}$.4459	$\overline{1}$.4488	$\overline{1}$.4517	$\overline{1}$.4546	5	10	15	20	24
16	$\overline{1}$.4575	$\overline{1}$.4603	$\overline{1}$.4632	$\overline{1}$.4660	$\overline{1}$.4688	$\overline{1}$.4716	$\overline{1}$.4744	$\overline{1}$.4771	$\overline{1}$.4799	$\overline{1}$.4826	5	9	14	19	23
17	$\overline{1}$.4853	$\overline{1}$.4880	$\overline{1}$.4907	$\overline{1}$.4934	$\overline{1}$.4961	$\overline{1}$.4987	$\overline{1}$.5014	$\overline{1}$.5040	$\overline{1}$.5066	$\overline{1}$.5092	4	9	13	18	22
18	$\overline{1}$.5118	$\overline{1}$.5143	$\overline{1}$.5169	$\overline{1}$.5195	$\overline{1}$.5220	$\overline{1}$.5245	$\overline{1}$.5270	$\overline{1}$.5295	$\overline{1}$.5320	$\overline{1}$.5345	4	8	13	17	21
19	$\overline{1}$.5370	$\overline{1}$.5394	$\overline{1}$.5419	$\overline{1}$.5443	$\overline{1}$.5467	$\overline{1}$.5491	$\overline{1}$.5516	$\overline{1}$.5539	$\overline{1}$.5563	$\overline{1}$.5587	4	8	12	16	20
20	$\overline{1}$.5611	$\overline{1}$.5634	$\overline{1}$.5658	$\overline{1}$.5681	$\overline{1}$.5704	$\overline{1}$.5727	$\overline{1}$.5750	$\overline{1}$.5773	$\overline{1}$.5796	$\overline{1}$.5819	4	8	12	15	19
21	$\overline{1}$.5842	$\overline{1}$.5864	$\overline{1}$.5887	$\overline{1}$.5909	$\overline{1}$.5932	$\overline{1}$.5954	$\overline{1}$.5976	$\overline{1}$.5998	$\overline{1}$.6020	$\overline{1}$.6042	4	7	11	15	18
22	$\overline{1}$.6064	$\overline{1}$.6086	$\overline{1}$.6108	$\overline{1}$.6129	$\overline{1}$.6151	$\overline{1}$.6172	$\overline{1}$.6194	$\overline{1}$.6215	$\overline{1}$.6236	$\overline{1}$.6257	4	7	11	14	18
23	$\overline{1}$.6279	$\overline{1}$.6300	$\overline{1}$.6321	$\overline{1}$.6341	$\overline{1}$.6362	$\overline{1}$.6383	$\overline{1}$.6404	$\overline{1}$.6424	$\overline{1}$.6445	$\overline{1}$.6465	3	7	10	14	17
24	$\overline{1}$.6486	$\overline{1}$.6506	$\overline{1}$.6527	$\overline{1}$.6547	$\overline{1}$.6567	$\overline{1}$.6587	$\overline{1}$.6607	$\overline{1}$.6627	$\overline{1}$.6647	$\overline{1}$.6667	3	7	10	13	17
25	$\overline{1}$.6687	$\overline{1}$.6706	$\overline{1}$.6726	$\overline{1}$.6746	$\overline{1}$.6765	$\overline{1}$.6785	$\overline{1}$.6804	$\overline{1}$.6824	$\overline{1}$.6843	$\overline{1}$.6863	3	6	10	13	16
26	$\overline{1}$.6882	$\overline{1}$.6901	$\overline{1}$.6920	$\overline{1}$.6939	$\overline{1}$.6958	$\overline{1}$.6977	$\overline{1}$.6996	$\overline{1}$.7015	$\overline{1}$.7034	$\overline{1}$.7053	3	6	10	13	16
27	$\overline{1}$.7072	$\overline{1}$.7090	$\overline{1}$.7109	$\overline{1}$.7128	$\overline{1}$.7146	$\overline{1}$.7165	$\overline{1}$.7183	$\overline{1}$.7202	$\overline{1}$.7220	$\overline{1}$.7238	3	6	9	12	15
28	$\overline{1}$.7257	$\overline{1}$.7275	$\overline{1}$.7293	$\overline{1}$.7311	$\overline{1}$.7330	$\overline{1}$.7348	$\overline{1}$.7366	$\overline{1}$.7384	$\overline{1}$.7402	$\overline{1}$.7420	3	6	9	12	15
29	$\overline{1}$.7438	$\overline{1}$.7455	$\overline{1}$.7473	$\overline{1}$.7491	$\overline{1}$.7509	$\overline{1}$.7526	$\overline{1}$.7544	$\overline{1}$.7562	$\overline{1}$.7579	$\overline{1}$.7597	3	6	9	12	15
30	$\overline{1}$.7614	$\overline{1}$.7632	$\overline{1}$.7649	$\overline{1}$.7667	$\overline{1}$.7684	$\overline{1}$.7701	$\overline{1}$.7719	$\overline{1}$.7736	$\overline{1}$.7753	$\overline{1}$.7771	3	6	9	12	15
31	$\overline{1}$.7788	$\overline{1}$.7805	$\overline{1}$.7822	$\overline{1}$.7839	$\overline{1}$.7856	$\overline{1}$.7873	$\overline{1}$.7890	$\overline{1}$.7907	$\overline{1}$.7924	$\overline{1}$.7941	3	6	9	11	14
32	$\overline{1}$.7958	$\overline{1}$.7975	$\overline{1}$.7992	$\overline{1}$.8008	$\overline{1}$.8025	$\overline{1}$.8042	$\overline{1}$.8059	$\overline{1}$.8075	$\overline{1}$.8092	$\overline{1}$.8109	3	6	8	11	14
33	$\overline{1}$.8125	$\overline{1}$.8142	$\overline{1}$.8158	$\overline{1}$.8175	$\overline{1}$.8191	$\overline{1}$.8208	$\overline{1}$.8224	$\overline{1}$.8241	$\overline{1}$.8257	$\overline{1}$.8274	3	6	8	11	14
34	$\overline{1}$.8290	$\overline{1}$.8306	$\overline{1}$.8323	$\overline{1}$.8339	$\overline{1}$.8355	$\overline{1}$.8371	$\overline{1}$.8388	$\overline{1}$.8404	$\overline{1}$.8420	$\overline{1}$.8436	3	5	8	11	14
35	$\overline{1}$.8452	$\overline{1}$.8468	$\overline{1}$.8484	$\overline{1}$.8501	$\overline{1}$.8517	$\overline{1}$.8533	$\overline{1}$.8549	$\overline{1}$.8565	$\overline{1}$.8581	$\overline{1}$.8597	3	5	8	11	13
36	$\overline{1}$.8613	$\overline{1}$.8629	$\overline{1}$.8644	$\overline{1}$.8660	$\overline{1}$.8676	$\overline{1}$.8692	$\overline{1}$.8708	$\overline{1}$.8724	$\overline{1}$.8740	$\overline{1}$.8755	3	5	8	11	13
37	$\overline{1}$.8771	$\overline{1}$.8787	$\overline{1}$.8803	$\overline{1}$.8818	$\overline{1}$.8834	$\overline{1}$.8850	$\overline{1}$.8865	$\overline{1}$.8881	$\overline{1}$.8897	$\overline{1}$.8912	3	5	8	10	13
38	$\overline{1}$.8928	$\overline{1}$.8944	$\overline{1}$.8959	$\overline{1}$.8975	$\overline{1}$.8990	$\overline{1}$.9006	$\overline{1}$.9022	$\overline{1}$.9037	$\overline{1}$.9053	$\overline{1}$.9068	3	5	8	10	13
39	$\overline{1}$.9084	$\overline{1}$.9099	$\overline{1}$.9115	$\overline{1}$.9130	$\overline{1}$.9146	$\overline{1}$.9161	$\overline{1}$.9176	$\overline{1}$.9192	$\overline{1}$.9207	$\overline{1}$.9223	3	5	8	10	13
40	$\overline{1}$.9238	$\overline{1}$.9254	$\overline{1}$.9269	$\overline{1}$.9284	$\overline{1}$.9300	$\overline{1}$.9315	$\overline{1}$.9330	$\overline{1}$.9346	$\overline{1}$.9361	$\overline{1}$.9376	3	5	8	10	13
41	$\overline{1}$.9392	$\overline{1}$.9407	$\overline{1}$.9422	$\overline{1}$.9438	$\overline{1}$.9453	$\overline{1}$.9468	$\overline{1}$.9483	$\overline{1}$.9499	$\overline{1}$.9514	$\overline{1}$.9529	3	5	8	10	13
42	$\overline{1}$.9544	$\overline{1}$.9560	$\overline{1}$.9575	$\overline{1}$.9590	$\overline{1}$.9605	$\overline{1}$.9621	$\overline{1}$.9636	$\overline{1}$.9651	$\overline{1}$.9666	$\overline{1}$.9681	3	5	8	10	13
43	$\overline{1}$.9697	$\overline{1}$.9712	$\overline{1}$.9727	$\overline{1}$.9742	$\overline{1}$.9757	$\overline{1}$.9772	$\overline{1}$.9788	$\overline{1}$.9803	$\overline{1}$.9818	$\overline{1}$.9833	3	5	8	10	13
44	$\overline{1}$.9848	$\overline{1}$.9864	$\overline{1}$.9879	$\overline{1}$.9894	$\overline{1}$.9909	$\overline{1}$.9924	$\overline{1}$.9939	$\overline{1}$.9955	$\overline{1}$.9970	$\overline{1}$.9985	3	5	8	10	13

To find $31.23 \times \tan 28° 37'$

$$\log(31.23 \times \tan 28° 37') = \log 31.23 + \log \tan 28° 37'$$
$$= 1.4946 + \overline{1}.7369$$
$$= 1.2315$$

From the anti-log tables the answer is 17.04.

Logarithms of Tangents

°	0' 0.0°	6' 0.1°	12' 0.2°	18' 0.3°	24' 0.4°	30' 0.5°	36' 0.6°	42' 0.7°	48' 0.8°	54' 0.9°	1'	2'	3'	4'	5'
45	0.0000	0.0015	0.0030	0.0045	0.0061	0.0076	0.0091	0.0106	0.0121	0.0136	3	5	8	10	13
46	0.0152	0.0167	0.0182	0.0197	0.0212	0.0228	0.0243	0.0258	0.0273	0.0288	3	5	8	10	13
47	0.0303	0.0319	0.0334	0.0349	0.0364	0.0379	0.0395	0.0410	0.0425	0.0440	3	5	8	10	13
48	0.0456	0.0471	0.0486	0.0501	0.0517	0.0532	0.0547	0.0562	0.0578	0.0593	3	5	8	10	13
49	0.0608	0.0624	0.0639	0.0654	0.0670	0.0685	0.0700	0.0716	0.0731	0.0746	3	5	8	10	13
50	0.0762	0.0777	0.0793	0.0808	0.0824	0.0839	0.0854	0.0870	0.0885	0.0901	3	5	8	10	13
51	0.0916	0.0932	0.0947	0.0963	0.0978	0.0994	0.1010	0.1025	0.1041	0.1056	3	5	8	10	13
52	0.1072	0.1088	0.1103	0.1119	0.1135	0.1150	0.1166	0.1182	0.1197	0.1213	3	5	8	10	13
53	0.1229	0.1245	0.1260	0.1276	0.1292	0.1308	0.1324	0.1340	0.1356	0.1371	3	5	8	11	13
54	0.1387	0.1403	0.1419	0.1435	0.1451	0.1467	0.1483	0.1499	0.1516	0.1532	3	5	8	11	13
55	0.1548	0.1564	0.1580	0.1596	0.1612	0.1629	0.1645	0.1661	0.1677	0.1694	3	5	8	11	14
56	0.1710	0.1726	0.1743	0.1759	0.1776	0.1792	0.1809	0.1825	0.1842	0.1858	3	6	8	11	14
57	0.1875	0.1891	0.1908	0.1925	0.1941	0.1958	0.1975	0.1992	0.2008	0.2025	3	6	8	11	14
58	0.2042	0.2059	0.2076	0.2093	0.2110	0.2127	0.2144	0.2161	0.2178	0.2195	3	6	9	11	14
59	0.2212	0.2229	0.2247	0.2264	0.2281	0.2299	0.2316	0.2333	0.2351	0.2368	3	6	9	12	15
60	0.2386	0.2403	0.2421	0.2438	0.2456	0.2474	0.2491	0.2509	0.2527	0.2545	3	6	9	12	15
61	0.2562	0.2580	0.2598	0.2616	0.2634	0.2652	0.2670	0.2689	0.2707	0.2725	3	6	9	12	15
62	0.2743	0.2762	0.2780	0.2798	0.2817	0.2835	0.2854	0.2872	0.2891	0.2910	3	6	9	12	15
63	0.2928	0.2947	0.2966	0.2985	0.3004	0.3023	0.3042	0.3061	0.3080	0.3099	3	6	9	13	16
64	0.3118	0.3137	0.3157	0.3176	0.3196	0.3215	0.3235	0.3254	0.3274	0.3294	3	6	10	13	16
65	0.3313	0.3333	0.3353	0.3373	0.3393	0.3413	0.3433	0.3453	0.3473	0.3494	3	7	10	13	17
66	0.3514	0.3535	0.3555	0.3576	0.3596	0.3617	0.3638	0.3659	0.3679	0.3700	3	7	10	14	17
67	0.3721	0.3743	0.3764	0.3785	0.3806	0.3828	0.3849	0.3871	0.3892	0.3914	4	7	11	14	18
68	0.3936	0.3958	0.3980	0.4002	0.4024	0.4046	0.4068	0.4091	0.4113	0.4136	4	7	11	15	18
69	0.4158	0.4181	0.4204	0.4227	0.4250	0.4273	0.4296	0.4319	0.4342	0.4366	4	8	12	15	19
70	0.4389	0.4413	0.4437	0.4461	0.4484	0.4509	0.4533	0.4557	0.4581	0.4606	4	8	12	16	20
71	0.4630	0.4655	0.4680	0.4705	0.4730	0.4755	0.4780	0.4805	0.4831	0.4857	4	8	13	17	21
72	0.4882	0.4908	0.4934	0.4960	0.4986	0.5013	0.5039	0.5066	0.5093	0.5120	4	9	13	18	22
73	0.5147	0.5174	0.5201	0.5229	0.5256	0.5284	0.5312	0.5340	0.5368	0.5397	5	9	14	19	23
74	0.5425	0.5454	0.5483	0.5512	0.5541	0.5570	0.5600	0.5629	0.5659	0.5689	5	10	15	20	24
75	0.5719	0.5750	0.5780	0.5811	0.5842	0.5873	0.5905	0.5936	0.5968	0.6000	5	10	16	21	26
76	0.6032	0.6065	0.6097	0.6130	0.6163	0.6196	0.6230	0.6264	0.6298	0.6332	6	11	17	22	28
77	0.6366	0.6401	0.6436	0.6471	0.6507	0.6542	0.6578	0.6615	0.6651	0.6688	6	12	18	24	30
78	0.6725	0.6763	0.6800	0.6838	0.6877	0.6915	0.6954	0.6994	0.7033	0.7073	6	13	19	26	32
79	0.7113	0.7154	0.7195	0.7236	0.7278	0.7320	0.7363	0.7406	0.7449	0.7493	7	14	21	28	35
80	0.7537	0.7581	0.7626	0.7672	0.7718	0.7764	0.7811	0.7858	0.7906	0.7954	8	16	23	31	39
81	0.8003	0.8052	0.8102	0.8152	0.8203	0.8255	0.8307	0.8360	0.8413	0.8467	9	17	26	35	43
82	0.8522	0.8577	0.8633	0.8690	0.8748	0.8806	0.8865	0.8924	0.8985	0.9046	10	20	29	39	49
83	0.9109	0.9172	0.9236	0.9301	0.9367	0.9433	0.9501	0.9570	0.9640	0.9711	11	22	34	45	56
84	0.9784	0.9857	0.9932	1.0008	1.0085	1.0164	1.0244	1.0326	1.0409	1.0494	13	26	40	53	66
85	1.0580	1.0669	1.0759	1.0850	1.0944	1.1040	1.1138	1.1238	1.1341	1.1446	16	32	48	64	81
86	1.1554	1.1664	1.1777	1.1893	1.2012	1.2135	1.2261	1.2391	1.2525	1.2663					
87	1.2806	1.2954	1.3106	1.3264	1.3429	1.3599	1.3777	1.3962	1.4155	1.4357		Differences			
88	1.4569	1.4792	1.5027	1.5275	1.5539	1.5819	1.6119	1.6441	1.6789	1.7167		untrustworthy			
89	1.7581	1.8038	1.8550	1.9130	1.9800	2.0591	2.1561	2.2810	2.4571	2.7581		here			

To find the angle A given that $\tan A = \dfrac{23.14}{17.68}$

$$\log \tan A = \log\left(\frac{23.14}{17.68}\right) = \log 23.14 - \log 17.68$$
$$= 1.3643 - 1.2475 = 0.1168$$

Using the log tangent tables

$$A = 52° 37'$$

Natural Sines

°	0' 0.0°	6' 0.1°	12' 0.2°	18' 0.3°	24' 0.4°	30' 0.5°	36' 0.6°	42' 0.7°	48' 0.8°	54' 0.9°	1'	2'	3'	4'	5'
0	0.0000	0.0017	0.0035	0.0052	0.0070	0.0087	0.0105	0.0122	0.0140	0.0157	3	6	9	12	15
1	0.0175	0.0192	0.0209	0.0227	0.0244	0.0262	0.0279	0.0297	0.0314	0.0332	3	6	9	12	15
2	0.0349	0.0366	0.0384	0.0401	0.0419	0.0436	0.0454	0.0471	0.0488	0.0506	3	6	9	12	15
3	0.0523	0.0541	0.0558	0.0576	0.0593	0.0610	0.0628	0.0645	0.0663	0.0680	3	6	9	12	15
4	0.0698	0.0715	0.0732	0.0750	0.0767	0.0785	0.0802	0.0819	0.0837	0.0854	3	6	9	12	14
5	0.0872	0.0889	0.0906	0.0924	0.0941	0.0958	0.0976	0.0993	0.1011	0.1028	3	6	9	12	14
6	0.1045	0.1063	0.1080	0.1097	0.1115	0.1132	0.1149	0.1167	0.1184	0.1201	3	6	9	12	14
7	0.1219	0.1236	0.1253	0.1271	0.1288	0.1305	0.1323	0.1340	0.1357	0.1374	3	6	9	12	14
8	0.1392	0.1409	0.1426	0.1444	0.1461	0.1478	0.1495	0.1513	0.1530	0.1547	3	6	9	11	14
9	0.1564	0.1582	0.1599	0.1616	0.1633	0.1650	0.1668	0.1685	0.1702	0.1719	3	6	9	11	14
10	0.1736	0.1754	0.1771	0.1788	0.1805	0.1822	0.1840	0.1857	0.1874	0.1891	3	6	9	11	14
11	0.1908	0.1925	0.1942	0.1959	0.1977	0.1994	0.2011	0.2028	0.2045	0.2062	3	6	9	11	14
12	0.2079	0.2096	0.2113	0.2130	0.2147	0.2164	0.2181	0.2198	0.2215	0.2232	3	6	9	11	14
13	0.2250	0.2267	0.2284	0.2300	0.2317	0.2334	0.2351	0.2368	0.2385	0.2402	3	6	8	11	14
14	0.2419	0.2436	0.2453	0.2470	0.2487	0.2504	0.2521	0.2538	0.2554	0.2571	3	6	8	11	14
15	0.2588	0.2605	0.2622	0.2639	0.2656	0.2672	0.2689	0.2706	0.2723	0.2740	3	6	8	11	14
16	0.2756	0.2773	0.2790	0.2807	0.2823	0.2840	0.2857	0.2874	0.2890	0.2907	3	6	8	11	14
17	0.2924	0.2940	0.2957	0.2974	0.2990	0.3007	0.3024	0.3040	0.3057	0.3074	3	6	8	11	14
18	0.3090	0.3107	0.3123	0.3140	0.3156	0.3173	0.3190	0.3206	0.3223	0.3239	3	6	8	11	14
19	0.3256	0.3272	0.3289	0.3305	0.3322	0.3338	0.3355	0.3371	0.3387	0.3404	3	5	8	11	14
20	0.3420	0.3437	0.3453	0.3469	0.3486	0.3502	0.3518	0.3535	0.3551	0.3567	3	5	8	11	14
21	0.3584	0.3600	0.3616	0.3633	0.3649	0.3665	0.3681	0.3697	0.3714	0.3730	3	5	8	11	14
22	0.3746	0.3762	0.3778	0.3795	0.3811	0.3827	0.3843	0.3859	0.3875	0.3891	3	5	8	11	13
23	0.3907	0.3923	0.3939	0.3955	0.3971	0.3987	0.4003	0.4019	0.4035	0.4051	3	5	8	11	13
24	0.4067	0.4083	0.4099	0.4115	0.4131	0.4147	0.4163	0.4179	0.4195	0.4210	3	5	8	11	13
25	0.4226	0.4242	0.4258	0.4274	0.4289	0.4305	0.4321	0.4337	0.4352	0.4368	3	5	8	11	13
26	0.4384	0.4399	0.4415	0.4431	0.4446	0.4462	0.4478	0.4493	0.4509	0.4524	3	5	8	10	13
27	0.4540	0.4555	0.4571	0.4586	0.4602	0.4617	0.4633	0.4648	0.4664	0.4679	3	5	8	10	13
28	0.4695	0.4710	0.4726	0.4741	0.4756	0.4772	0.4787	0.4802	0.4818	0.4833	3	5	8	10	13
29	0.4848	0.4863	0.4879	0.4894	0.4909	0.4924	0.4939	0.4955	0.4970	0.4985	3	5	8	10	13
30	0.5000	0.5015	0.5030	0.5045	0.5060	0.5075	0.5090	0.5105	0.5120	0.5135	3	5	8	10	13
31	0.5150	0.5165	0.5180	0.5195	0.5210	0.5225	0.5240	0.5255	0.5270	0.5284	2	5	7	10	12
32	0.5299	0.5314	0.5329	0.5344	0.5358	0.5373	0.5388	0.5402	0.5417	0.5432	2	5	7	10	12
33	0.5446	0.5461	0.5476	0.5490	0.5505	0.5519	0.5534	0.5548	0.5563	0.5577	2	5	7	10	12
34	0.5592	0.5606	0.5621	0.5635	0.5650	0.5664	0.5678	0.5693	0.5707	0.5721	2	5	7	10	12
35	0.5736	0.5750	0.5764	0.5779	0.5793	0.5807	0.5821	0.5835	0.5850	0.5864	2	5	7	9	12
36	0.5878	0.5892	0.5906	0.5920	0.5934	0.5948	0.5962	0.5976	0.5990	0.6004	2	5	7	9	12
37	0.6018	0.6032	0.6046	0.6060	0.6074	0.6088	0.6101	0.6115	0.6129	0.6143	2	5	7	9	12
38	0.6157	0.6170	0.6184	0.6198	0.6211	0.6225	0.6239	0.6252	0.6266	0.6280	2	5	7	9	11
39	0.6293	0.6307	0.6320	0.6334	0.6347	0.6361	0.6374	0.6388	0.6401	0.6414	2	4	7	9	11
40	0.6428	0.6441	0.6455	0.6468	0.6481	0.6494	0.6508	0.6521	0.6534	0.6547	2	4	7	9	11
41	0.6561	0.6574	0.6587	0.6600	0.6613	0.6626	0.6639	0.6652	0.6665	0.6678	2	4	7	9	11
42	0.6691	0.6704	0.6717	0.6730	0.6743	0.6756	0.6769	0.6782	0.6794	0.6807	2	4	6	9	11
43	0.6820	0.6833	0.6845	0.6858	0.6871	0.6884	0.6896	0.6909	0.6921	0.6934	2	4	6	8	11
44	0.6947	0.6959	0.6972	0.6984	0.6997	0.7009	0.7022	0.7034	0.7046	0.7059	2	4	6	8	10

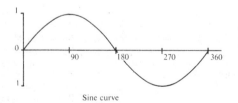

Sine curve

Natural Sines

°	0' 0.0°	6' 0.1°	12' 0.2°	18' 0.3°	24' 0.4°	30' 0.5°	36' 0.6°	42' 0.7°	48' 0.8°	54' 0.9°	1'	2'	3'	4'	5'
45	0.7071	0.7083	0.7096	0.7108	0.7120	0.7133	0.7145	0.7157	0.7169	0.7181	2	4	6	8	10
46	0.7193	0.7206	0.7218	0.7230	0.7242	0.7254	0.7266	0.7278	0.7290	0.7302	2	4	6	8	10
47	0.7314	0.7325	0.7337	0.7349	0.7361	0.7373	0.7385	0.7396	0.7408	0.7420	2	4	6	8	10
48	0.7431	0.7443	0.7455	0.7466	0.7478	0.7490	0.7501	0.7513	0.7524	0.7536	2	4	6	8	10
49	0.7547	0.7558	0.7570	0.7581	0.7593	0.7604	0.7615	0.7627	0.7638	0.7649	2	4	6	8	9
50	0.7660	0.7672	0.7683	0.7694	0.7705	0.7716	0.7727	0.7738	0.7749	0.7760	2	4	6	7	9
51	0.7771	0.7782	0.7793	0.7804	0.7815	0.7826	0.7837	0.7848	0.7859	0.7869	2	4	5	7	9
52	0.7880	0.7891	0.7902	0.7912	0.7923	0.7934	0.7944	0.7955	0.7965	0.7976	2	4	5	7	9
53	0.7986	0.7997	0.8007	0.8018	0.8028	0.8039	0.8049	0.8059	0.8070	0.8080	2	3	5	7	9
54	0.8090	0.8100	0.8111	0.8121	0.8131	0.8141	0.8151	0.8161	0.8171	0.8181	2	3	5	7	8
55	0.8192	0.8202	0.8211	0.8221	0.8231	0.8241	0.8251	0.8261	0.8271	0.8281	2	3	5	7	8
56	0.8290	0.8300	0.8310	0.8320	0.8329	0.8339	0.8348	0.8358	0.8368	0.8377	2	3	5	6	8
57	0.8387	0.8396	0.8406	0.8415	0.8425	0.8434	0.8443	0.8453	0.8462	0.8471	2	3	5	6	8
58	0.8480	0.8490	0.8499	0.8508	0.8517	0.8526	0.8536	0.8545	0.8554	0.8563	2	3	5	6	8
59	0.8572	0.8581	0.8590	0.8599	0.8607	0.8616	0.8625	0.8634	0.8643	0.8652	1	3	4	6	7
60	0.8660	0.8669	0.8678	0.8686	0.8695	0.8704	0.8712	0.8721	0.8729	0.8738	1	3	4	6	7
61	0.8746	0.8755	0.8763	0.8771	0.8780	0.8788	0.8796	0.8805	0.8813	0.8821	1	3	4	6	7
62	0.8829	0.8838	0.8846	0.8854	0.8862	0.8870	0.8878	0.8886	0.8894	0.8902	1	3	4	5	7
63	0.8910	0.8918	0.8926	0.8934	0.8942	0.8949	0.8957	0.8965	0.8973	0.8980	1	3	4	5	6
64	0.8988	0.8996	0.9003	0.9011	0.9018	0.9026	0.9033	0.9041	0.9048	0.9056	1	3	4	5	6
65	0.9063	0.9070	0.9078	0.9085	0.9092	0.9100	0.9107	0.9114	0.9121	0.9128	1	2	4	5	6
66	0.9135	0.9143	0.9150	0.9157	0.9164	0.9171	0.9178	0.9184	0.9191	0.9198	1	2	3	5	6
67	0.9205	0.9212	0.9219	0.9225	0.9232	0.9239	0.9245	0.9252	0.9259	0.9265	1	2	3	4	6
68	0.9272	0.9278	0.9285	0.9291	0.9298	0.9304	0.9311	0.9317	0.9323	0.9330	1	2	3	4	5
69	0.9336	0.9342	0.9348	0.9354	0.9361	0.9367	0.9373	0.9379	0.9385	0.9391	1	2	3	4	5
70	0.9397	0.9403	0.9409	0.9415	0.9421	0.9426	0.9432	0.9438	0.9444	0.9449	1	2	3	4	5
71	0.9455	0.9461	0.9466	0.9472	0.9478	0.9483	0.9489	0.9494	0.9500	0.9505	1	2	3	4	5
72	0.9511	0.9516	0.9521	0.9527	0.9532	0.9537	0.9542	0.9548	0.9553	0.9558	1	2	3	3	4
73	0.9563	0.9568	0.9573	0.9578	0.9583	0.9588	0.9593	0.9598	0.9603	0.9608	1	2	2	3	4
74	0.9613	0.9617	0.9622	0.9627	0.9632	0.9636	0.9641	0.9646	0.9650	0.9655	1	2	2	3	4
75	0.9659	0.9664	0.9668	0.9673	0.9677	0.9681	0.9686	0.9690	0.9694	0.9699	1	1	2	3	4
76	0.9703	0.9707	0.9711	0.9715	0.9720	0.9724	0.9728	0.9732	0.9736	0.9740	1	1	2	3	3
77	0.9744	0.9748	0.9751	0.9755	0.9759	0.9763	0.9767	0.9770	0.9774	0.9778	1	1	2	2	3
78	0.9781	0.9785	0.9789	0.9792	0.9796	0.9799	0.9803	0.9806	0.9810	0.9813	1	1	2	2	3
79	0.9816	0.9820	0.9823	0.9826	0.9829	0.9833	0.9836	0.9839	0.9842	0.9845	1	1	2	2	3
80	0.9848	0.9851	0.9854	0.9857	0.9860	0.9863	0.9866	0.9869	0.9871	0.9874	0	1	1	2	2
81	0.9877	0.9880	0.9882	0.9885	0.9888	0.9890	0.9893	0.9895	0.9898	0.9900	0	1	1	2	2
82	0.9903	0.9905	0.9907	0.9910	0.9912	0.9914	0.9917	0.9919	0.9921	0.9923	0	1	1	1	2
83	0.9925	0.9928	0.9930	0.9932	0.9934	0.9936	0.9938	0.9940	0.9942	0.9943	0	1	1	1	2
84	0.9945	0.9947	0.9949	0.9951	0.9952	0.9954	0.9956	0.9957	0.9959	0.9960	0	1	1	1	1
85	0.9962	0.9963	0.9965	0.9966	0.9968	0.9969	0.9971	0.9972	0.9973	0.9974	0	0	1	1	1
86	0.9976	0.9977	0.9978	0.9979	0.9980	0.9981	0.9982	0.9983	0.9984	0.9985	0	0	1	1	1
87	0.9986	0.9987	0.9988	0.9989	0.9990	0.9990	0.9991	0.9992	0.9993	0.9993	0	0	0	1	1
88	0.9994	0.9995	0.9995	0.9996	0.9996	0.9997	0.9997	0.9997	0.9998	0.9998	0	0	0	0	0
89	0.9998	0.9999	0.9999	0.9999	0.9999	1.0000	1.0000	1.0000	1.0000	1.0000	0	0	0	0	0
90	1.0000														

Quadrant	Angle	sin A =	Examples
first	0 to 90°	sin A	sin 34°38' = 0.5683
second	90° to 180°	sin(180°−A)	sin 145°22' = sin(180°−145°22')
third	180° to 270°	−sin(A−180°)	= sin 34°38' = 0.5683
fourth	270° to 360°	−sin(360°−A)	sin 214°38' = −sin(214°38'−180°)
			= −sin 34°38' = −0.5683
			sin 325°22' = −sin(360°−325°22')
			= −sin 34°38' = −0.5683

Natural Cosines

Numbers in difference columns to be *subtracted*, not added.

°	0' 0.0°	6' 0.1°	12' 0.2°	18' 0.3°	24' 0.4°	30' 0.5°	36' 0.6°	42' 0.7°	48' 0.8°	54' 0.9°	1'	2'	3'	4'	5'
0	1.0000	1.0000	1.0000	1.0000	1.0000	1.0000	0.9999	0.9999	0.9999	0.9999	0	0	0	0	0
1	0.9998	0.9998	0.9998	0.9997	0.9997	0.9997	0.9996	0.9996	0.9995	0.9995	0	0	0	0	0
2	0.9994	0.9993	0.9993	0.9992	0.9991	0.9990	0.9990	0.9989	0.9988	0.9987	0	0	0	1	1
3	0.9986	0.9985	0.9984	0.9983	0.9982	0.9981	0.9980	0.9979	0.9978	0.9977	0	0	1	1	1
4	0.9976	0.9974	0.9973	0.9972	0.9971	0.9969	0.9968	0.9966	0.9965	0.9963	0	0	1	1	1
5	0.9962	0.9960	0.9959	0.9957	0.9956	0.9954	0.9952	0.9951	0.9949	0.9947	0	1	1	1	1
6	0.9945	0.9943	0.9942	0.9940	0.9938	0.9936	0.9934	0.9932	0.9930	0.9928	0	1	1	1	2
7	0.9925	0.9923	0.9921	0.9919	0.9917	0.9914	0.9912	0.9910	0.9907	0.9905	0	1	1	1	2
8	0.9903	0.9900	0.9898	0.9895	0.9893	0.9890	0.9888	0.9885	0.9882	0.9880	0	1	1	2	2
9	0.9877	0.9874	0.9871	0.9869	0.9866	0.9863	0.9860	0.9857	0.9854	0.9851	0	1	1	2	2
10	0.9848	0.9845	0.9842	0.9839	0.9836	0.9833	0.9829	0.9826	0.9823	0.9820	1	1	2	2	3
11	0.9816	0.9813	0.9810	0.9806	0.9803	0.9799	0.9796	0.9792	0.9789	0.9785	1	1	2	2	3
12	0.9781	0.9778	0.9774	0.9770	0.9767	0.9763	0.9759	0.9755	0.9751	0.9748	1	1	2	2	3
13	0.9744	0.9740	0.9736	0.9732	0.9728	0.9724	0.9720	0.9715	0.9711	0.9707	1	1	2	3	3
14	0.9703	0.9699	0.9694	0.9690	0.9686	0.9681	0.9677	0.9673	0.9668	0.9664	1	1	2	3	4
15	0.9659	0.9655	0.9650	0.9646	0.9641	0.9636	0.9632	0.9627	0.9622	0.9617	1	2	2	3	4
16	0.9613	0.9608	0.9603	0.9598	0.9593	0.9588	0.9583	0.9578	0.9573	0.9568	1	2	2	3	4
17	0.9563	0.9558	0.9553	0.9548	0.9542	0.9537	0.9532	0.9527	0.9521	0.9516	1	2	3	3	4
18	0.9511	0.9505	0.9500	0.9494	0.9489	0.9483	0.9478	0.9472	0.9466	0.9461	1	2	3	4	5
19	0.9455	0.9449	0.9444	0.9438	0.9432	0.9426	0.9421	0.9415	0.9409	0.9403	1	2	3	4	5
20	0.9397	0.9391	0.9385	0.9379	0.9373	0.9367	0.9361	0.9354	0.9348	0.9342	1	2	3	4	5
21	0.9336	0.9330	0.9323	0.9317	0.9311	0.9304	0.9298	0.9291	0.9285	0.9278	1	2	3	4	5
22	0.9272	0.9265	0.9259	0.9252	0.9245	0.9239	0.9232	0.9225	0.9219	0.9212	1	2	3	4	6
23	0.9205	0.9198	0.9191	0.9184	0.9178	0.9171	0.9164	0.9157	0.9150	0.9143	1	2	3	5	6
24	0.9135	0.9128	0.9121	0.9114	0.9107	0.9100	0.9092	0.9085	0.9078	0.9070	1	2	4	5	6
25	0.9063	0.9056	0.9048	0.9041	0.9033	0.9026	0.9018	0.9011	0.9003	0.8996	1	3	4	5	6
26	0.8988	0.8980	0.8973	0.8965	0.8957	0.8949	0.8942	0.8934	0.8926	0.8918	1	3	4	5	6
27	0.8910	0.8902	0.8894	0.8886	0.8878	0.8870	0.8862	0.8854	0.8846	0.8838	1	3	4	5	7
28	0.8829	0.8821	0.8813	0.8805	0.8796	0.8788	0.8780	0.8771	0.8763	0.8755	1	3	4	6	7
29	0.8746	0.8738	0.8729	0.8721	0.8712	0.8704	0.8695	0.8686	0.8678	0.8669	1	3	4	6	7
30	0.8660	0.8652	0.8643	0.8634	0.8625	0.8616	0.8607	0.8599	0.8590	0.8581	1	3	4	6	7
31	0.8572	0.8563	0.8554	0.8545	0.8536	0.8526	0.8517	0.8508	0.8499	0.8490	2	3	5	6	8
32	0.8480	0.8471	0.8462	0.8453	0.8443	0.8434	0.8425	0.8415	0.8406	0.8396	2	3	5	6	8
33	0.8387	0.8377	0.8368	0.8358	0.8348	0.8339	0.8329	0.8320	0.8310	0.8300	2	3	5	6	8
34	0.8290	0.8281	0.8271	0.8261	0.8251	0.8241	0.8231	0.8221	0.8211	0.8202	2	3	5	7	8
35	0.8192	0.8181	0.8171	0.8161	0.8151	0.8141	0.8131	0.8121	0.8111	0.8100	2	3	5	7	8
36	0.8090	0.8080	0.8070	0.8059	0.8049	0.8039	0.8028	0.8018	0.8007	0.7997	2	3	5	7	9
37	0.7986	0.7976	0.7965	0.7955	0.7944	0.7934	0.7923	0.7912	0.7902	0.7891	2	4	5	7	9
38	0.7880	0.7869	0.7859	0.7848	0.7837	0.7826	0.7815	0.7804	0.7793	0.7782	2	4	5	7	9
39	0.7771	0.7760	0.7749	0.7738	0.7727	0.7716	0.7705	0.7694	0.7683	0.7672	2	4	6	7	9
40	0.7660	0.7649	0.7638	0.7627	0.7615	0.7604	0.7593	0.7581	0.7570	0.7559	2	4	6	8	9
41	0.7547	0.7536	0.7524	0.7513	0.7501	0.7490	0.7478	0.7466	0.7455	0.7443	2	4	6	8	10
42	0.7431	0.7420	0.7408	0.7396	0.7385	0.7373	0.7361	0.7349	0.7337	0.7325	2	4	6	8	10
43	0.7314	0.7302	0.7290	0.7278	0.7266	0.7254	0.7242	0.7230	0.7218	0.7206	2	4	6	8	10
44	0.7193	0.7181	0.7169	0.7157	0.7145	0.7133	0.7120	0.7108	0.7096	0.7083	2	4	6	8	10

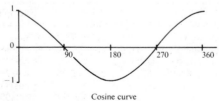

Cosine curve

Natural Cosines

Numbers in difference columns to be *subtracted*, not added.

°	0' 0.0°	6' 0.1°	12' 0.2°	18' 0.3°	24' 0.4°	30' 0.5°	36' 0.6°	42' 0.7°	48' 0.8°	54' 0.9°	1'	2'	3'	4'	5'
45	0.7071	0.7059	0.7046	0.7034	0.7022	0.7009	0.6997	0.6984	0.6972	0.6959	2	4	6	8	10
46	0.6947	0.6934	0.6921	0.6909	0.6896	0.6884	0.6871	0.6858	0.6845	0.6833	2	4	6	8	11
47	0.6820	0.6807	0.6794	0.6782	0.6769	0.6756	0.6743	0.6730	0.6717	0.6704	2	4	6	9	11
48	0.6691	0.6678	0.6665	0.6652	0.6639	0.6626	0.6613	0.6600	0.6587	0.6574	2	4	7	9	11
49	0.6561	0.6547	0.6534	0.6521	0.6508	0.6494	0.6481	0.6468	0.6455	0.6441	2	4	7	9	11
50	0.6428	0.6414	0.6401	0.6388	0.6374	0.6361	0.6347	0.6334	0.6320	0.6307	2	4	7	9	11
51	0.6293	0.6280	0.6266	0.6252	0.6239	0.6225	0.6211	0.6198	0.6184	0.6170	2	5	7	9	11
52	0.6157	0.6143	0.6129	0.6115	0.6101	0.6088	0.6074	0.6060	0.6046	0.6032	2	5	7	9	12
53	0.6018	0.6004	0.5990	0.5976	0.5962	0.5948	0.5934	0.5920	0.5906	0.5892	2	5	7	9	12
54	0.5878	0.5864	0.5850	0.5835	0.5821	0.5807	0.5793	0.5779	0.5764	0.5750	2	5	7	9	12
55	0.5736	0.5721	0.5707	0.5693	0.5678	0.5664	0.5650	0.5635	0.5621	0.5606	2	5	7	10	12
56	0.5592	0.5577	0.5563	0.5548	0.5534	0.5519	0.5505	0.5490	0.5476	0.5461	2	5	7	10	12
57	0.5446	0.5432	0.5417	0.5402	0.5388	0.5373	0.5358	0.5344	0.5329	0.5314	2	5	7	10	12
58	0.5299	0.5284	0.5270	0.5255	0.5240	0.5225	0.5210	0.5195	0.5180	0.5165	2	5	7	10	12
59	0.5150	0.5135	0.5120	0.5105	0.5090	0.5075	0.5060	0.5045	0.5030	0.5015	3	5	8	10	13
60	0.5000	0.4985	0.4970	0.4955	0.4939	0.4924	0.4909	0.4894	0.4879	0.4863	3	5	8	10	13
61	0.4848	0.4833	0.4818	0.4802	0.4787	0.4772	0.4756	0.4741	0.4726	0.4710	3	5	8	10	13
62	0.4695	0.4679	0.4664	0.4648	0.4633	0.4617	0.4602	0.4586	0.4571	0.4555	3	5	8	10	13
63	0.4540	0.4524	0.4509	0.4493	0.4478	0.4462	0.4446	0.4431	0.4415	0.4399	3	5	8	10	13
64	0.4384	0.4368	0.4352	0.4337	0.4321	0.4305	0.4289	0.4274	0.4258	0.4242	3	5	8	11	13
65	0.4226	0.4210	0.4195	0.4179	0.4163	0.4147	0.4131	0.4115	0.4099	0.4083	3	5	8	11	13
66	0.4067	0.4051	0.4035	0.4019	0.4003	0.3987	0.3971	0.3955	0.3939	0.3923	3	5	8	11	13
67	0.3907	0.3891	0.3875	0.3859	0.3843	0.3827	0.3811	0.3795	0.3778	0.3762	3	5	8	11	13
68	0.3746	0.3730	0.3714	0.3697	0.3681	0.3665	0.3649	0.3633	0.3616	0.3600	3	5	8	11	14
69	0.3584	0.3567	0.3551	0.3535	0.3518	0.3502	0.3486	0.3469	0.3453	0.3437	3	5	8	11	14
70	0.3420	0.3404	0.3387	0.3371	0.3355	0.3338	0.3322	0.3305	0.3289	0.3272	3	5	8	11	14
71	0.3256	0.3239	0.3223	0.3206	0.3190	0.3173	0.3156	0.3140	0.3123	0.3107	3	6	8	11	14
72	0.3090	0.3074	0.3057	0.3040	0.3024	0.3007	0.2990	0.2974	0.2957	0.2940	3	6	8	11	14
73	0.2924	0.2907	0.2890	0.2874	0.2857	0.2840	0.2823	0.2807	0.2790	0.2773	3	6	8	11	14
74	0.2756	0.2740	0.2723	0.2706	0.2689	0.2672	0.2656	0.2639	0.2622	0.2605	3	6	8	11	14
75	0.2588	0.2571	0.2554	0.2538	0.2521	0.2504	0.2487	0.2470	0.2453	0.2436	3	6	8	11	14
76	0.2419	0.2402	0.2385	0.2368	0.2351	0.2334	0.2317	0.2300	0.2284	0.2267	3	6	8	11	14
77	0.2250	0.2233	0.2215	0.2198	0.2181	0.2164	0.2147	0.2130	0.2113	0.2096	3	6	9	11	14
78	0.2079	0.2062	0.2045	0.2028	0.2011	0.1994	0.1977	0.1959	0.1942	0.1925	3	6	9	11	14
79	0.1908	0.1891	0.1874	0.1857	0.1840	0.1822	0.1805	0.1788	0.1771	0.1754	3	6	9	11	14
80	0.1736	0.1719	0.1702	0.1685	0.1668	0.1650	0.1633	0.1616	0.1599	0.1582	3	6	9	11	14
81	0.1564	0.1547	0.1530	0.1513	0.1495	0.1478	0.1461	0.1444	0.1426	0.1409	3	6	9	11	14
82	0.1392	0.1374	0.1357	0.1340	0.1323	0.1305	0.1288	0.1271	0.1253	0.1236	3	6	9	12	14
83	0.1219	0.1201	0.1184	0.1167	0.1149	0.1132	0.1115	0.1097	0.1080	0.1063	3	6	9	12	14
84	0.1045	0.1028	0.1011	0.0993	0.0976	0.0958	0.0941	0.0924	0.0906	0.0889	3	6	9	12	14
85	0.0872	0.0854	0.0837	0.0819	0.0802	0.0785	0.0767	0.0750	0.0732	0.0715	3	6	9	12	14
86	0.0698	0.0680	0.0663	0.0645	0.0628	0.0610	0.0593	0.0576	0.0558	0.0541	3	6	9	12	15
87	0.0523	0.0506	0.0488	0.0471	0.0454	0.0436	0.0419	0.0401	0.0384	0.0366	3	6	9	12	15
88	0.0349	0.0332	0.0314	0.0297	0.0279	0.0262	0.0244	0.0227	0.0209	0.0192	3	6	9	12	15
89	0.0175	0.0157	0.0140	0.0122	0.0105	0.0087	0.0070	0.0052	0.0035	0.0017	3	6	9	12	15
90	0.0000														

Quadrant	Angle	cos A =	Examples
first	0 to 90°	cos A	cos 33°26' = 0.8345
second	90° to 180°	−cos(180°−A)	cos 146°34' = −cos(180°−146°34')
third	180° to 270°	−cos(A−180°)	= −cos 33°26' = −0.8345
fourth	270° to 360°	cos(360°−A)	cos 213°26' = −cos(213°26'−180°)
			= −cos 33°26' = −0.8345
			cos 326°34' = cos(360°−326°34')
			= cos 33°26' = 0.8345

Natural Tangents

°	0' 0.0°	6' 0.1°	12' 0.2°	18' 0.3°	24' 0.4°	30' 0.5°	36' 0.6°	42' 0.7°	48' 0.8°	54' 0.9°	1'	2'	3'	4'	5'
0	0.0000	0.0017	0.0035	0.0052	0.0070	0.0087	0.0105	0.0122	0.0140	0.0157	3	6	9	12	15
1	0.0175	0.0192	0.0209	0.0227	0.0244	0.0262	0.0279	0.0297	0.0314	0.0332	3	6	9	12	15
2	0.0349	0.0367	0.0384	0.0402	0.0419	0.0437	0.0454	0.0472	0.0489	0.0507	3	6	9	12	15
3	0.0524	0.0542	0.0559	0.0577	0.0594	0.0612	0.0629	0.0647	0.0664	0.0682	3	6	9	12	15
4	0.0699	0.0717	0.0734	0.0752	0.0769	0.0787	0.0805	0.0822	0.0840	0.0857	3	6	9	12	15
5	0.0875	0.0892	0.0910	0.0928	0.0945	0.0963	0.0981	0.0998	0.1016	0.1033	3	6	9	12	15
6	0.1051	0.1069	0.1086	0.1104	0.1122	0.1139	0.1157	0.1175	0.1192	0.1210	3	6	9	12	15
7	0.1228	0.1246	0.1263	0.1281	0.1299	0.1317	0.1334	0.1352	0.1370	0.1388	3	6	9	12	15
8	0.1405	0.1423	0.1441	0.1459	0.1477	0.1495	0.1512	0.1530	0.1548	0.1566	3	6	9	12	15
9	0.1584	0.1602	0.1620	0.1638	0.1655	0.1673	0.1691	0.1709	0.1727	0.1745	3	6	9	12	15
10	0.1763	0.1781	0.1799	0.1817	0.1835	0.1853	0.1871	0.1890	0.1908	0.1926	3	6	9	12	15
11	0.1944	0.1962	0.1980	0.1998	0.2016	0.2035	0.2053	0.2071	0.2089	0.2107	3	6	9	12	15
12	0.2126	0.2144	0.2162	0.2180	0.2199	0.2217	0.2235	0.2254	0.2272	0.2290	3	6	9	12	15
13	0.2309	0.2327	0.2345	0.2364	0.2382	0.2401	0.2419	0.2438	0.2456	0.2475	3	6	9	12	15
14	0.2493	0.2512	0.2530	0.2549	0.2568	0.2586	0.2605	0.2623	0.2642	0.2661	3	6	9	12	16
15	0.2679	0.2698	0.2717	0.2736	0.2754	0.2773	0.2792	0.2811	0.2830	0.2849	3	6	9	13	16
16	0.2867	0.2886	0.2905	0.2924	0.2943	0.2962	0.2981	0.3000	0.3019	0.3038	3	6	9	13	16
17	0.3057	0.3076	0.3096	0.3115	0.3134	0.3153	0.3172	0.3191	0.3211	0.3230	3	6	10	13	16
18	0.3249	0.3269	0.3288	0.3307	0.3327	0.3346	0.3365	0.3385	0.3404	0.3424	3	6	10	13	16
19	0.3443	0.3463	0.3482	0.3502	0.3522	0.3541	0.3561	0.3581	0.3600	0.3620	3	7	10	13	16
20	0.3640	0.3659	0.3679	0.3699	0.3719	0.3739	0.3759	0.3779	0.3799	0.3819	3	7	10	13	17
21	0.3839	0.3859	0.3879	0.3899	0.3919	0.3939	0.3959	0.3979	0.4000	0.4020	3	7	10	13	17
22	0.4040	0.4061	0.4081	0.4101	0.4122	0.4142	0.4163	0.4183	0.4204	0.4224	3	7	10	14	17
23	0.4245	0.4265	0.4286	0.4307	0.4327	0.4348	0.4369	0.4390	0.4411	0.4431	3	7	10	14	17
24	0.4452	0.4473	0.4494	0.4515	0.4536	0.4557	0.4578	0.4599	0.4621	0.4642	4	7	11	14	18
25	0.4663	0.4684	0.4706	0.4727	0.4748	0.4770	0.4791	0.4813	0.4834	0.4856	4	7	11	14	18
26	0.4877	0.4899	0.4921	0.4942	0.4964	0.4986	0.5008	0.5029	0.5051	0.5073	4	7	11	15	18
27	0.5095	0.5117	0.5139	0.5161	0.5184	0.5206	0.5228	0.5250	0.5272	0.5295	4	7	11	15	18
28	0.5317	0.5340	0.5362	0.5384	0.5407	0.5430	0.5452	0.5475	0.5498	0.5520	4	8	11	15	19
29	0.5543	0.5566	0.5589	0.5612	0.5635	0.5658	0.5681	0.5704	0.5727	0.5750	4	8	12	15	19
30	0.5774	0.5797	0.5820	0.5844	0.5867	0.5890	0.5914	0.5938	0.5961	0.5985	4	8	12	16	20
31	0.6009	0.6032	0.6056	0.6080	0.6104	0.6128	0.6152	0.6176	0.6200	0.6224	4	8	12	16	20
32	0.6249	0.6273	0.6297	0.6322	0.6346	0.6371	0.6395	0.6420	0.6445	0.6469	4	8	12	16	20
33	0.6494	0.6519	0.6544	0.6569	0.6594	0.6619	0.6644	0.6669	0.6694	0.6720	4	8	13	17	21
34	0.6745	0.6771	0.6796	0.6822	0.6847	0.6873	0.6899	0.6924	0.6950	0.6976	4	9	13	17	21
35	0.7002	0.7028	0.7054	0.7080	0.7107	0.7133	0.7159	0.7186	0.7212	0.7239	4	9	13	17	22
36	0.7265	0.7292	0.7319	0.7346	0.7373	0.7400	0.7427	0.7454	0.7481	0.7508	5	9	14	18	23
37	0.7536	0.7563	0.7590	0.7618	0.7646	0.7673	0.7701	0.7729	0.7757	0.7785	5	9	14	18	23
38	0.7813	0.7841	0.7869	0.7898	0.7926	0.7954	0.7983	0.8012	0.8040	0.8069	5	9	14	19	24
39	0.8098	0.8127	0.8156	0.8185	0.8214	0.8243	0.8273	0.8302	0.8332	0.8361	5	10	15	20	24
40	0.8391	0.8421	0.8451	0.8481	0.8511	0.8541	0.8571	0.8601	0.8632	0.8662	5	10	15	20	25
41	0.8693	0.8724	0.8754	0.8785	0.8816	0.8847	0.8878	0.8910	0.8941	0.8972	5	10	16	21	26
42	0.9004	0.9036	0.9067	0.9099	0.9131	0.9163	0.9195	0.9228	0.9260	0.9293	5	11	16	21	27
43	0.9325	0.9358	0.9391	0.9424	0.9457	0.9490	0.9523	0.9556	0.9590	0.9623	6	11	17	22	28
44	0.9657	0.9691	0.9725	0.9759	0.9793	0.9827	0.9861	0.9896	0.9930	0.9965	6	11	17	23	28

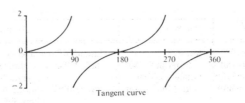

Tangent curve

Natural Tangents

°	0' 0.0°	6' 0.1°	12' 0.2°	18' 0.3°	24' 0.4°	30' 0.5°	36' 0.6°	42' 0.7°	48' 0.8°	54' 0.9°	1'	2'	3'	4'	5'
45	1.0000	1.0035	1.0070	1.0105	1.0141	1.0176	1.0212	1.0247	1.0283	1.0319	6	12	18	24	30
46	1.0355	1.0392	1.0428	1.0464	1.0501	1.0538	1.0575	1.0612	1.0649	1.0686	6	12	18	25	31
47	1.0724	1.0761	1.0799	1.0837	1.0875	1.0913	1.0951	1.0990	1.1028	1.1067	6	13	19	25	32
48	1.1106	1.1145	1.1184	1.1224	1.1263	1.1303	1.1343	1.1383	1.1423	1.1463	7	13	20	27	33
49	1.1504	1.1544	1.1585	1.1626	1.1667	1.1708	1.1750	1.1792	1.1833	1.1875	7	14	21	28	34
50	1.1918	1.1960	1.2002	1.2045	1.2088	1.2131	1.2174	1.2218	1.2261	1.2305	7	14	22	29	36
51	1.2349	1.2393	1.2437	1.2482	1.2527	1.2572	1.2617	1.2662	1.2708	1.2753	8	15	23	30	38
52	1.2799	1.2846	1.2892	1.2938	1.2985	1.3032	1.3079	1.3127	1.3175	1.3222	8	16	24	31	39
53	1.3270	1.3319	1.3367	1.3416	1.3465	1.3514	1.3564	1.3613	1.3663	1.3713	8	16	25	33	41
54	1.3764	1.3814	1.3865	1.3916	1.3968	1.4019	1.4071	1.4124	1.4176	1.4229	9	17	26	34	43
55	1.4281	1.4335	1.4388	1.4442	1.4496	1.4550	1.4605	1.4659	1.4715	1.4770	9	18	27	36	45
56	1.4826	1.4882	1.4938	1.4994	1.5051	1.5108	1.5166	1.5224	1.5282	1.5340	10	19	29	38	48
57	1.5399	1.5458	1.5517	1.5577	1.5637	1.5697	1.5757	1.5818	1.5880	1.5941	10	20	30	40	50
58	1.6003	1.6066	1.6128	1.6191	1.6255	1.6319	1.6383	1.6447	1.6512	1.6577	11	21	32	43	53
59	1.6643	1.6709	1.6775	1.6842	1.6909	1.6977	1.7045	1.7113	1.7182	1.7251	11	23	34	45	56
60	1.7321	1.7391	1.7461	1.7532	1.7603	1.7675	1.7747	1.7820	1.7893	1.7966	12	24	36	48	60
61	1.8040	1.8115	1.8190	1.8265	1.8341	1.8418	1.8495	1.8572	1.8650	1.8728	13	26	38	51	64
62	1.8807	1.8887	1.8967	1.9047	1.9128	1.9210	1.9292	1.9375	1.9458	1.9542	14	27	41	55	68
63	1.9626	1.9711	1.9797	1.9883	1.9970	2.0057	2.0145	2.0233	2.0323	2.0413	15	29	44	58	73
64	2.0503	2.0594	2.0686	2.0778	2.0872	2.0965	2.1060	2.1155	2.1251	2.1348	16	31	47	63	78
65	2.1445	2.1543	2.1642	2.1742	2.1842	2.1943	2.2045	2.2148	2.2251	2.2355	17	34	51	68	85
66	2.2460	2.2566	2.2673	2.2781	2.2889	2.2998	2.3109	2.3220	2.3332	2.3445	18	37	55	73	92
67	2.3559	2.3673	2.3789	2.3906	2.4023	2.4142	2.4262	2.4383	2.4504	2.4627	20	40	60	79	99
68	2.4751	2.4876	2.5002	2.5129	2.5257	2.5386	2.5517	2.5649	2.5782	2.5916	22	43	65	87	108
69	2.6051	2.6187	2.6325	2.6464	2.6605	2.6746	2.6889	2.7034	2.7179	2.7326	24	47	71	95	119
70	2.7475	2.7625	2.7776	2.7929	2.8083	2.8239	2.8397	2.8556	2.8716	2.8878	26	52	78	104	131
71	2.9042	2.9208	2.9375	2.9544	2.9714	2.9887	3.0061	3.0237	3.0415	3.0595	29	58	87	116	145
72	3.0777	3.0961	3.1146	3.1334	3.1524	3.1716	3.1910	3.2106	3.2205	3.2506	32	64	96	129	161
73	3.2709	3.2914	3.3122	3.3332	3.3544	3.3759	3.3977	3.4197	3.4420	3.4646	36	72	108	144	180
74	3.4874	3.5105	3.5339	3.5576	3.5816	3.6059	3.6305	3.6554	3.6806	3.7062	41	81	122	163	204
75	3.7321	3.7583	3.7848	3.8118	3.8391	3.8667	3.8947	3.9232	3.9520	3.9812	46	93	139	186	232
76	4.0108	4.0408	4.0713	4.1022	4.1335	4.1653	4.1976	4.2303	4.2635	4.2972	53	107	160	213	267
77	4.3315	4.3662	4.4015	4.4374	4.4737	4.5107	4.5483	4.5864	4.6252	4.6646					
78	4.7046	4.7453	4.7867	4.8288	4.8716	4.9152	4.9594	5.0045	5.0504	5.0970					
79	5.1446	5.1929	5.2422	5.2924	5.3435	5.3955	5.4486	5.5026	5.5578	5.6140					
80	5.6713	5.7297	5.7894	5.8502	5.9124	5.9758	6.0405	6.1066	6.1742	6.2432					
81	6.3138	6.3859	6.4596	6.5350	6.6122	6.6912	6.7720	6.8548	6.9395	7.0264					
82	7.1154	7.2066	7.3002	6.3962	7.4947	7.5958	7.6996	7.8062	7.9158	8.0285		Differences			
83	8.1443	8.2636	8.3863	8.5126	8.6427	8.7769	8.9152	9.0579	9.2052	9.3572		untrustworthy			
84	9.5144	9.677	9.845	10.02	10.20	10.39	10.58	10.78	10.99	11.20		here			
85	11.43	11.66	11.91	12.16	12.43	12.71	13.00	13.30	13.62	13.95					
86	14.30	14.67	15.06	15.46	15.89	16.35	16.83	17.34	17.89	18.46					
87	19.08	19.74	20.45	21.20	22.02	22.90	23.86	24.90	26.03	27.27					
88	28.64	30.14	31.82	33.69	35.80	38.19	40.92	44.07	47.74	52.08					
89	57.29	63.66	71.62	81.85	95.49	114.6	143.2	191.0	286.5	573.0					
90	∞														

Quadrant	Angle	tan A =	Examples
first	0 to 90°	tan A	tan 56°17′ = 1.4986
second	90° to 180°	−tan(180°−A)	tan 123°43′ = −tan(180°−123°43′)
third	180° to 270°	tan(A−180°)	= −tan 56°17′ = −1.4986
fourth	270° to 360°	−tan(360°−A)	tan 236°17′ = tan(236°17′−180°)
			= tan 56°17′ = 1.4986
			tan 303°43′ = −tan(360°−303°43′)
			= −tan 56°17′ = −1.4986

Natural Cotangents

Numbers in difference columns to be *subtracted*, not added.

°	0' 0.0°	6' 0.1°	12' 0.2°	18' 0.3°	24' 0.4°	30' 0.5°	36' 0.6°	42' 0.7°	48' 0.8°	54' 0.9°					
0	∞	573.0	286.5	191.0	143.2	114.6	95.49	81.85	71.62	63.66					
1	57.29	52.08	47.74	44.07	40.92	38.19	35.80	33.69	31.82	30.14		Differences			
2	28.64	27.27	26.03	24.90	23.86	22.90	22.02	21.20	20.45	19.74		untrustworthy			
3	19.08	18.46	17.89	17.34	16.83	16.35	15.89	15.46	15.06	14.67		here			
4	14.30	13.95	13.62	13.30	13.00	12.71	12.43	12.16	11.91	11.66					
5	11.43	11.20	10.99	10.78	10.58	10.39	10.20	10.02	9.84	9.68					
6	9.514	9.357	9.205	9.058	8.915	8.777	8.643	8.513	8.386	8.264					
7	8.144	8.028	7.916	7.806	7.700	7.596	7.495	7.396	7.300	7.207					
8	7.115	7.026	6.940	6.855	6.772	6.691	6.612	6.535	6.460	6.386					
9	6.314	6.243	6.174	6.107	6.041	5.976	5.912	5.850	5.789	5.730	1'	2'	3'	4'	5'
10	5.6713	5.6140	5.5578	5.5026	5.4486	5.3955	5.3435	5.2924	5.2422	5.1929	87	175	263	350	438
11	5.1446	5.0970	5.0504	5.0045	4.9594	4.9152	4.8716	4.8288	4.7867	4.7453	73	146	220	293	366
12	4.7046	4.6646	4.6252	4.5864	4.5483	4.5107	4.4737	4.4373	4.4015	4.3662	62	124	186	248	310
13	4.3315	4.2972	4.2635	4.2303	4.1976	4.1653	4.1335	4.1022	4.0713	4.0408	53	107	160	214	267
14	4.0108	3.9812	3.9520	3.9232	3.8947	3.8667	3.8391	3.8118	3.7848	3.7583	46	93	139	186	232
15	3.7321	3.7062	3.6806	3.6554	3.6305	3.6059	3.5816	3.5576	3.5339	3.5105	41	81	122	163	203
16	3.4874	3.4646	3.4420	3.4197	3.3977	3.3759	3.3544	3.3332	3.3122	3.2914	36	72	108	144	180
17	3.2709	3.2506	3.2305	3.2106	3.1910	3.1716	3.1524	3.1334	3.1146	3.0961	32	64	97	129	161
18	3.0777	3.0595	3.0415	3.0237	3.0061	2.9887	2.9714	2.9544	2.9375	2.9208	29	58	87	116	144
19	2.9042	2.8878	2.8716	2.8556	2.8397	2.8239	2.8083	2.7929	2.7776	2.7625	26	52	78	104	130
20	2.7475	2.7326	2.7179	2.7034	2.6889	2.6746	2.6605	2.6464	2.6325	2.6187	24	47	71	95	119
21	2.6051	2.5916	2.5782	2.5649	2.5517	2.5386	2.5257	2.5129	2.5002	2.4876	22	43	65	87	108
22	2.4751	2.4627	2.4504	2.4383	2.4262	2.4142	2.4023	2.3906	2.3789	2.3673	20	40	60	79	99
23	2.3559	2.3445	2.3332	2.3220	2.3109	2.2998	2.2889	2.2781	2.2673	2.2566	18	37	55	73	91
24	2.2460	2.2355	2.2251	2.2148	2.2045	2.1943	2.1842	2.1742	2.1642	2.1543	17	34	51	68	85
25	2.1445	2.1348	2.1251	2.1155	2.1060	2.0965	2.0872	2.0778	2.0686	2.0594	16	31	47	63	78
26	2.0503	2.0413	2.0323	2.0233	2.0145	2.0057	1.9970	1.9883	1.9797	1.9711	15	29	44	58	73
27	1.9626	1.9542	1.9458	1.9375	1.9292	1.9210	1.9128	1.9047	1.8967	1.8887	14	27	41	55	68
28	1.8807	1.8728	1.8650	1.8572	1.8495	1.8418	1.8341	1.8265	1.8190	1.8115	13	26	38	51	64
29	1.8040	1.7966	1.7893	1.7820	1.7747	1.7675	1.7603	1.7532	1.7461	1.7391	12	24	36	48	60
30	1.7321	1.7251	1.7182	1.7113	1.7045	1.6977	1.6909	1.6842	1.6775	1.6709	11	23	34	45	56
31	1.6643	1.6577	1.6512	1.6447	1.6383	1.6319	1.6255	1.6191	1.6128	1.6066	11	21	32	43	53
32	1.6003	1.5941	1.5880	1.5818	1.5757	1.5697	1.5637	1.5577	1.5517	1.5458	10	20	30	40	50
33	1.5399	1.5340	1.5282	1.5224	1.5166	1.5108	1.5051	1.4994	1.4938	1.4882	10	19	29	38	48
34	1.4826	1.4770	1.4715	1.4659	1.4605	1.4550	1.4496	1.4442	1.4388	1.4335	9	18	27	36	45
35	1.4281	1.4229	1.4176	1.4124	1.4071	1.4019	1.3968	1.3916	1.3865	1.3814	9	17	26	34	43
36	1.3764	1.3713	1.3663	1.3613	1.3564	1.3514	1.3465	1.3416	1.3367	1.3319	8	16	25	33	41
37	1.3270	1.3222	1.3175	1.3127	1.3079	1.3032	1.2985	1.2938	1.2892	1.2846	8	16	24	31	39
38	1.2799	1.2753	1.2708	1.2662	1.2617	1.2572	1.2527	1.2482	1.2437	1.2393	8	15	23	30	38
39	1.2349	1.2305	1.2261	1.2218	1.2174	1.2131	1.2088	1.2045	1.2002	1.1960	7	14	22	29	36
40	1.1918	1.1875	1.1833	1.1792	1.1750	1.1708	1.1667	1.1626	1.1585	1.1544	7	14	21	28	34
41	1.1504	1.1463	1.1423	1.1383	1.1343	1.1303	1.1263	1.1224	1.1184	1.1145	7	13	20	27	33
42	1.1106	1.1067	1.1028	1.0990	1.0951	1.0913	1.0875	1.0837	1.0799	1.0761	6	13	19	25	32
43	1.0724	1.0686	1.0649	1.0612	1.0575	1.0538	1.0501	1.0464	1.0428	1.0392	6	12	18	25	31
44	1.0355	1.0319	1.0283	1.0247	1.0212	1.0176	1.0141	1.0105	1.0070	1.0035	6	12	18	24	30

$$\cot A = \frac{1}{\tan A}$$

To find $\dfrac{1}{\tan 30° 22'}$ look up $\cot 30° 22' = 1.7068$

Natural Cotangents

Numbers in difference columns to be *subtracted*, not added.

°	0' 0.0°	6' 0.1°	12' 0.2°	18' 0.3°	24' 0.4°	30' 0.5°	36' 0.6°	42' 0.7°	48' 0.8°	54' 0.9°	1'	2'	3'	4'	5'
45	1.0000	0.9965	0.9930	0.9896	0.9861	0.9827	0.9793	0.9759	0.9725	0.9691	6	11	17	23	29
46	0.9657	0.9623	0.9590	0.9556	0.9523	0.9490	0.9457	0.9424	0.9391	0.9358	6	11	17	22	28
47	0.9325	0.9293	0.9260	0.9228	0.9195	0.9163	0.9131	0.9099	0.9067	0.9036	5	11	16	21	27
48	0.9004	0.8972	0.8941	0.8910	0.8878	0.8847	0.8816	0.8785	0.8754	0.8724	5	10	16	21	26
49	0.8693	0.8662	0.8632	0.8601	0.8571	0.8541	0.8511	0.8481	0.8451	0.8421	5	10	15	20	25
50	0.8391	0.8361	0.8332	0.8302	0.8273	0.8243	0.8214	0.8185	0.8156	0.8127	5	10	15	20	24
51	0.8098	8.8069	0.8040	0.8012	0.7983	0.7954	0.7926	0.7898	0.7869	0.7841	5	9	14	19	24
52	0.7813	0.7785	0.7757	0.7729	0.7701	0.7673	0.7646	0.7618	0.7590	0.7563	5	9	14	18	23
53	0.7536	0.7508	0.7481	0.7454	0.7427	0.7400	0.7373	0.7346	0.7319	0.7292	5	9	14	18	23
54	0.7265	0.7239	0.7212	0.7186	0.7159	0.7133	0.7107	0.7080	0.7054	0.7028	4	9	13	18	22
55	0.7002	0.6976	0.6950	0.6924	0.6899	0.6873	0.6847	0.6822	0.6796	0.6771	4	9	13	17	21
56	0.6745	0.6720	0.6694	0.6669	0.6644	0.6619	0.6594	0.6569	0.6544	0.6519	4	8	13	17	21
57	0.6494	0.6469	0.6445	0.6420	0.6395	0.6371	0.6346	0.6322	0.6297	0.6273	4	8	12	16	20
58	0.6249	0.6224	0.6200	0.6176	0.6152	0.6128	0.6104	0.6080	0.6056	0.6032	4	8	12	16	20
59	0.6009	0.5985	0.5961	0.5938	0.5914	0.5890	0.5867	0.5844	0.5820	0.5797	4	8	12	16	20
60	0.5774	0.5750	0.5727	0.5704	0.5681	0.5658	0.5635	0.5612	0.5589	0.5566	4	8	12	15	19
61	0.5543	0.5520	0.5498	0.5475	0.5452	0.5430	0.5407	0.5384	0.5362	0.5340	4	8	11	15	19
62	0.5317	0.5295	0.5272	0.5250	0.5228	0.5206	0.5184	0.5161	0.5139	0.5117	4	7	11	15	18
63	0.5095	0.5073	0.5051	0.5029	0.5008	0.4986	0.4964	0.4942	0.4921	0.4899	4	7	11	15	18
64	0.4877	0.4856	0.4834	0.4813	0.4791	0.4770	0.4748	0.4727	0.4706	0.4684	4	7	11	14	18
65	0.4663	0.4642	0.4621	0.4599	0.4578	0.4557	0.4536	0.4515	0.4494	0.4473	4	7	11	14	18
66	0.4452	0.4431	0.4411	0.4390	0.4369	0.4348	0.4327	0.4307	0.4286	0.4265	3	7	10	14	17
67	0.4245	0.4224	0.4204	0.4183	0.4163	0.4142	0.4122	0.4101	0.4081	0.4061	3	7	10	14	17
68	0.4040	0.4020	0.4000	0.3979	0.3959	0.3939	0.3919	0.3899	0.3879	0.3859	3	7	10	13	17
69	0.3839	0.3819	0.3799	0.3779	0.3759	0.3739	0.3719	0.3699	0.3679	0.3659	3	7	10	13	17
70	0.3640	0.3620	0.3600	0.3581	0.3561	0.3541	0.3522	0.3502	0.3482	0.3463	3	7	10	13	16
71	0.3443	0.3424	0.3404	0.3385	0.3365	0.3346	0.3327	0.3307	0.3288	0.3269	3	6	10	13	16
72	0.3249	0.3230	0.3211	0.3191	0.3172	0.3153	0.3134	0.3115	0.3096	0.3076	3	6	10	13	16
73	0.3057	0.3038	0.3019	0.3000	0.2981	0.2962	0.2943	0.2924	0.2905	0.2886	3	6	9	13	16
74	0.2867	0.2849	0.2830	0.2811	0.2792	0.2773	0.2754	0.2736	0.2717	0.2698	3	6	9	13	16
75	0.2679	0.2661	0.2642	0.2623	0.2605	0.2586	0.2568	0.2549	0.2530	0.2512	3	6	9	12	16
76	0.2493	0.2475	0.2456	0.2438	0.2419	0.2401	0.2382	0.2364	0.2345	0.2327	3	6	9	12	15
77	0.2309	0.2290	0.2272	0.2254	0.2235	0.2217	0.2199	0.2180	0.2162	0.2144	3	6	9	12	15
78	0.2126	0.2107	0.2089	0.2071	0.2053	0.2035	0.2016	0.1998	0.1980	0.1962	3	6	9	12	15
79	0.1944	0.1926	0.1908	0.1890	0.1871	0.1853	0.1835	0.1817	0.1799	0.1781	3	6	9	12	15
80	0.1763	0.1745	0.1727	0.1709	0.1691	0.1673	0.1655	0.1638	0.1620	0.1602	3	6	9	12	15
81	0.1584	0.1566	0.1548	0.1530	0.1512	0.1495	0.1477	0.1459	0.1441	0.1423	3	6	9	12	15
82	0.1405	0.1388	0.1370	0.1352	0.1334	0.1317	0.1299	0.1281	0.1263	0.1246	3	6	9	12	15
83	0.1228	0.1210	0.1192	0.1175	0.1157	0.1139	0.1122	0.1104	0.1086	0.1069	3	6	9	12	15
84	0.1051	0.1033	0.1016	0.0998	0.0981	0.0963	0.0945	0.0928	0.0910	0.0892	3	6	9	12	15
85	0.0875	0.0857	0.0840	0.0822	0.0805	0.0787	0.0769	0.0752	0.0734	0.0717	3	6	9	12	15
86	0.0699	0.0682	0.0664	0.0647	0.0629	0.0612	0.0594	0.0577	0.0559	0.0542	3	6	9	12	15
87	0.0524	0.0507	0.0489	0.0472	0.0454	0.0437	0.0419	0.0402	0.0384	0.0367	3	6	9	12	15
88	0.0349	0.0332	0.0314	0.0297	0.0279	0.0262	0.0244	0.0227	0.0209	0.0192	3	6	9	12	15
89	0.0175	0.0157	0.0140	0.0122	0.0105	0.0087	0.0070	0.0052	0.0035	0.0017	3	6	9	12	15

Quadrant	Angle	cot A =	Examples
first	0°–90°	cot A	cot 56° 17' = 0.6673
second	90°–180°	−cot(180°−A)	cot 123° 43' = −cot(180°−123° 43')
third	180°–270°	cot(A−180°)	= −cot 56° 17' = −0.6673
fourth	270°–360°	−tan(360°−A)	cot 236° 17' = cot(236° 17'−180°)
			= cot 56° 17' = 0.6673
			cot 303° 43' = −cot(360°−303° 43')
			= −cot 56° 17' = −0.6673

Natural Secants

°	0' 0.0°	6' 0.1°	12' 0.2°	18' 0.3°	24' 0.4°	30' 0.5°	36' 0.6°	42' 0.7°	48' 0.8°	54' 0.9°	1'	2'	3'	4'	5'
0	1.0000	1.0000	1.0000	1.0000	1.0000	1.0000	1.0001	1.0001	1.0001	1.0001					
1	1.0002	1.0002	1.0002	1.0003	1.0003	1.0003	1.0004	1.0004	1.0005	1.0006					
2	1.0006	1.0007	1.0007	1.0008	1.0009	1.0010	1.0010	1.0011	1.0012	1.0013					
3	1.0014	1.0015	1.0016	1.0017	1.0018	1.0019	1.0020	1.0021	1.0022	1.0023	0	0	0	1	1
4	1.0024	1.0026	1.0027	1.0028	1.0030	1.0031	1.0032	1.0034	1.0035	1.0037	0	0	1	1	1
5	1.0038	1.0040	1.0041	1.0043	1.0045	1.0046	1.0048	1.0050	1.0051	1.0053	0	1	1	1	1
6	1.0055	1.0057	1.0059	1.0061	1.0063	1.0065	1.0067	1.0069	1.0071	1.0073	0	1	1	1	2
7	1.0075	1.0077	1.0079	1.0082	1.0084	1.0086	1.0089	1.0091	1.0093	1.0096	0	1	1	2	2
8	1.0098	1.0101	1.0103	1.0106	1.0108	1.0111	1.0114	1.0116	1.0119	1.0122	0	1	1	2	2
9	1.0125	1.0127	1.0130	1.0133	1.0136	1.0139	1.0142	1.0145	1.0148	1.0151	0	1	1	2	2
10	1.0154	1.0157	1.0161	1.0164	1.0167	1.0170	1.0174	1.0177	1.0180	0.0184	1	1	2	2	3
11	1.0187	1.0191	1.0194	1.0198	1.0201	1.0205	1.0209	1.0212	1.0216	1.0220	1	1	2	2	3
12	1.0223	1.0227	1.0231	1.0235	1.0239	1.0243	1.0247	1.0251	1.0255	1.0259	1	1	2	3	3
13	1.0263	1.0267	1.0271	1.0276	1.0280	1.0284	1.0288	1.0293	1.0297	1.0302	1	1	2	3	4
14	1.0306	1.0311	1.0315	1.0320	1.0324	1.0329	1.0334	1.0338	1.0343	1.0348	1	2	2	3	4
15	1.0353	1.0358	1.0363	1.0367	1.0372	1.0377	1.0382	1.0388	1.0393	1.0398	1	2	3	3	4
16	1.0403	1.0408	1.0413	1.0419	1.0424	1.0429	1.0435	1.0440	1.0446	1.0451†	1	2	3	4	4
17	1.0457	1.0463	1.0468	1.0474	1.0480	1.0485	1.0491	1.0497	1.0503	1.0509	1	2	3	4	5
18	1.0515	1.0521	1.0527	1.0533	1.0539	1.0545	1.0551	1.0557	1.0564	1.0570	1	2	3	4	5
19	1.0576	1.0583	1.0589	1.0595	1.0602	1.0608	1.0615	1.0622	1.0628	1.0635	1	2	3	4	5
20	1.0642	1.0649	1.0655	1.0662	1.0669	1.0676	1.0683	1.0690	1.0697	1.0704	1	2	3	5	6
21	1.0711	1.0719	1.0726	1.0733	1.0740	1.0748	1.0755	1.0763	1.0770	1.0778	1	2	4	5	6
22	1.0785	1.0793	1.0801	1.0808	1.0816	1.0824	1.0832	1.0840	1.0848	1.0856	1	3	4	5	7
23	1.0864	1.0872	1.0880	1.0888	1.0896	1.0904	1.0913	1.0921	1.0929	1.0938	1	3	4	6	7
24	1.0946	1.0955	1.0963	1.0972	1.0981	1.0989	1.0998	1.1007	1.1016	1.1025	1	3	4	6	7
25	1.1034	1.1043	1.1052	1.1061	1.1070	1.1079	1.1089	1.1098	1.1107	1.1117	2	3	5	6	8
26	1.1126	1.1136	1.1145	1.1155	1.1164	1.1174	1.1184	1.1194	1.1203	1.1213	2	3	5	6	8
27	1.1223	1.1233	1.1243	1.1253	1.1264	1.1274	1.1284	1.1294	1.1305	1.1315	2	3	5	7	9
28	1.1326	1.1336	1.1347	1.1357	1.1368	1.1379	1.1390	1.1401	1.1412	1.1423	2	4	5	7	9
29	1.1434	1.1445	1.1456	1.1467	1.1478	1.1490	1.1501	1.1512	1.1524	1.1535	2	4	6	8	9
30	1.1547	1.1559	1.1570	1.1582	1.1594	1.1606	1.1618	1.1630	1.1642	1.1654	2	4	6	8	10
31	1.1666	1.1679	1.1691	1.1703	1.1716	1.1728	1.1741	1.1753	1.1766	1.1779	2	4	6	8	10
32	1.1792	1.1805	1.1818	1.1831	1.1844	1.1857	1.1870	1.1883	1.1897	1.1910	2	4	7	9	11
33	1.1924	1.1937	1.1951	1.1964	1.1978	1.1992	1.2006	1.2020	1.2034	1.2048	2	5	7	9	12
34	1.2062	1.2076	1.2091	1.2105	1.2120	1.2134	1.2149	1.2163	1.2178	1.2193	2	5	7	10	12
35	1.2208	1.2223	1.2238	1.2253	1.2268	1.2283	1.2299	1.2314	1.2329	1.2345	3	5	8	10	13
36	1.2361	1.2376	1.2392	1.2408	1.2424	1.2440	1.2456	1.2472	1.2489	1.2505	3	5	8	11	13
37	1.2521	1.2538	1.2554	1.2571	1.2588	1.2605	1.2622	1.2639	1.2656	1.2673	3	6	8	11	14
38	1.2690	1.2708	1.2725	1.2742	1.2760	1.2778	1.2796	1.2813	1.2831	1.2849	3	6	9	12	15
39	1.2868	1.2886	1.2904	1.2923	1.2941	1.2960	1.2978	1.2997	1.3016	1.3035	3	6	9	12	16
40	1.3054	1.3073	1.3093	1.3112	1.3131	1.3151	1.3171	1.3190	1.3210	1.3230	3	7	10	13	16
41	1.3250	1.3270	1.3291	1.3311	1.3331	1.3352	1.3373	1.3393	1.3414	1.3435	3	7	10	14	17
42	1.3456	1.3478	1.3499	1.3520	1.3542	1.3563	1.3585	1.3607	1.3629	1.3651	4	7	11	14	18
43	1.3673	1.3696	1.3718	1.3741	1.3763	1.3786	1.3809	1.3832	1.3855	1.3878	4	8	11	15	19
44	1.3902	1.3925	1.3949	1.3972	1.3996	1.4020	1.4044	1.4069	1.4093	1.4118	4	8	12	16	20

$$\sec A = \frac{1}{\cos A}$$

To find $\dfrac{1}{\cos 35° 40'}$ look up $\sec 35° 40' = 1.2309$

Natural Secants

°	0' 0.0°	6' 0.1°	12' 0.2°	18' 0.3°	24' 0.4°	30' 0.5°	36' 0.6°	42' 0.7°	48' 0.8°	54' 0.9°	1'	2'	3'	4'	5'
45	1.4142	1.4167	1.4192	1.4217	1.4242	1.4267	1.4293	1.4318	1.4344	1.4370	4	8	13	17	21
46	1.4396	1.4422	1.4448	1.4474	1.4501	1.4527	1.4554	1.4581	1.4608	1.4635	4	9	13	18	22
47	1.4663	1.4690	1.4718	1.4746	1.4774	1.4802	1.4830	1.4859	1.4887	1.4916	5	9	14	19	23
48	1.4945	1.4974	1.5003	1.5032	1.5062	1.5092	1.5121	1.5151	1.5182	1.5212	5	10	15	20	25
49	1.5243	1.5273	1.5304	1.5335	1.5366	1.5398	1.5429	1.5461	1.5493	1.5525	5	10	16	21	26
50	1.5557	1.5590	1.5622	1.5655	1.5688	1.5721	1.5755	1.5788	1.5822	1.5856	6	11	17	22	28
51	1.5890	1.5925	1.5959	1.5994	1.6029	1.6064	1.6099	1.6135	1.6171	1.6207	6	12	18	23	29
52	1.6243	1.6279	1.6316	1.6353	1.6390	1.6427	1.6464	1.6502	1.6540	1.6578	6	12	19	25	31
53	1.6616	1.6655	1.6694	1.6733	1.6772	1.6812	1.6852	1.6892	1.6932	1.6972	7	13	20	26	33
54	1.7013	1.7054	1.7095	1.7137	1.7179	1.7221	1.7263	1.7305	1.7348	1.7391	7	14	21	28	35
55	1.7434	1.7478	1.7522	1.7566	1.7610	1.7655	1.7700	1.7745	1.7791	1.7837	7	15	22	30	37
56	1.7883	1.7929	1.7976	1.8023	1.8070	1.8118	1.8166	1.8214	1.8263	1.8312	8	16	24	32	40
57	1.8361	1.8410	1.8460	1.8510	1.8561	1.8612	1.8663	1.8714	1.8766	1.8818	8	17	25	34	42
58	1.8871	1.8924	1.8977	1.9031	1.9084	1.9139	1.9194	1.9249	1.9304	1.9360	9	18	27	36	45
59	1.9416	1.9473	1.9530	1.9587	1.9645	1.9703	1.9762	1.9821	1.9880	1.9940	10	19	29	39	49
60	2.0000	2.0061	2.0122	2.0183	2.0245	2.0308	2.0371	2.0434	2.0498	2.0562	10	21	31	42	52
61	2.0627	2.0692	2.0757	2.0824	2.0890	2.0957	2.1025	2.1093	2.1162	2.1231	11	22	34	45	56
62	2.1301	2.1371	2.1441	2.1513	2.1584	2.1657	2.1730	2.1803	2.1877	2.1952	12	24	36	48	61
63	2.2027	2.2103	2.2179	2.2256	2.2333	2.2412	2.2490	2.2570	2.2650	2.2730	13	26	39	52	65
64	2.2812	2.2894	2.2976	2.3060	2.3144	2.3228	2.3314	2.3400	2.3486	2.3574	14	28	43	57	71
65	2.3662	2.3751	2.3841	2.3931	2.4022	2.4114	2.4207	2.4300	2.4395	2.4490	15	31	46	62	77
66	2.4586	2.4683	2.4780	2.4879	2.4978	2.5078	2.5180	2.5282	2.5384	2.5488	17	34	50	67	84
67	2.5593	2.5699	2.5805	2.5913	2.6022	2.6131	2.6242	2.6354	2.6466	2.6580	18	37	55	73	92
68	2.6695	2.6811	2.6927	2.7046	2.7165	2.7285	2.7407	2.7529	2.7653	2.7778	20	40	60	81	101
69	2.7904	2.8032	2.8161	2.8291	2.8422	2.8555	2.8688	2.8824	2.8960	2.9099	22	44	67	89	111
70	2.9238	2.9379	2.9521	2.9665	2.9811	2.9957	3.0106	3.0256	3.0407	3.0561	25	49	74	98	123
71	3.0716	3.0872	3.1030	3.1190	3.1352	3.1515	3.1681	3.1848	3.2017	3.2188	27	55	82	110	137
72	3.2361	3.2535	3.2712	3.2891	3.3072	3.3255	3.3440	3.3628	3.3817	3.4009	31	61	92	123	153
73	3.4203	3.4399	3.4598	3.4799	3.5003	3.5209	3.5418	3.5629	3.5843	3.6060	35	69	104	138	173
74	3.6280	3.6502	3.6727	3.6955	3.7186	3.7420	3.7657	3.7897	3.8140	3.8387	39	79	118	157	196
75	3.8637	3.8890	3.9147	3.9408	3.9672	3.9939	4.0211	4.0486	4.0765	4.1048	45	90	135	180	225
76	4.1336	4.1627	4.1923	4.2223	4.2527	4.2837	4.3150	4.3469	4.3792	4.4121	52	104	156	207	260
77	4.4454	4.4793	4.5137	4.5486	4.5841	4.6202	4.6569	4.6942	4.7321	4.7706	61	121	182	242	303
78	4.8097	4.8496	4.8901	4.9313	4.9732	5.0159	5.0593	5.1034	5.1484	5.1942	72	143	215	287	359
79	5.2408	5.2883	5.3367	5.3860	5.4362	5.4874	5.5396	5.5928	5.6470	5.7023	86	172	258	344	431
80	5.759	5.816	5.875	5.935	5.996	6.059	6.123	6.188	6.255	6.323					
81	6.392	6.464	6.537	6.611	6.687	6.765	6.845	6.927	7.011	7.097					
82	7.185	7.276	7.368	7.463	7.561	7.661	7.764	7.870	7.979	8.091					
83	8.206	8.324	8.446	8.571	8.700	8.834	8.971	9.113	9.259	9.411					
84	9.57	9.73	9.90	10.07	10.25	10.43	10.63	10.83	11.03	11.25		Differences untrustworthy here			
85	11.47	11.71	11.95	12.20	12.47	12.75	13.03	13.34	13.65	13.99					
86	14.34	14.70	15.09	15.50	15.93	16.38	16.86	17.37	17.91	18.49					
87	19.11	19.77	20.47	21.23	22.04	22.93	23.88	24.92	26.05	27.29					
88	28.65	30.16	31.84	33.71	35.81	38.20	40.93	44.08	47.75	52.09					
89	57.30	63.66	71.62	81.85	95.49	114.6	143.2	191.0	286.5	573.0					

Quadrant	Angle	sec A	Examples
first	0–90°	sec A	sec 33° 26' = 1.1983
second	90°–180°	−sec(180°−A)	sec 146° 34' = −sec(180°−146° 34')
third	180°–270°	−sec(A−180°)	= −sec 33° 26' = +1.1983
fourth	270°–360°	sec(360°−A)	sec 213° 26' = −sec(213° 26'−180°)
			= −sec 33° 26' = −1.1983
			sec 326° 34' = sec(360°−326° 34')
			= sec 33° 26' = 1.1983

Natural Cosecants

Numbers in difference columns to be *subtracted*, not added.

°	0' 0.0°	6' 0.1°	12' 0.2°	18' 0.3°	24' 0.4°	30' 0.5°	36' 0.6°	42' 0.7°	48' 0.8°	54' 0.9°	1'	2'	3'	4'	5'
0	∞	573.0	286.5	191.0	143.2	114.6	95.49	81.85	71.62	63.66					
1	57.30	52.09	47.75	44.08	40.93	38.20	35.81	33.71	31.84	30.16		Differences			
2	28.65	27.29	26.05	24.92	23.88	22.93	22.04	21.23	20.47	19.77		untrustworthy			
3	19.11	18.49	17.91	17.37	16.86	16.38	15.93	15.50	15.09	14.70		here			
4	14.34	13.99	13.65	13.34	13.03	12.75	12.47	12.20	11.95	11.71					
5	11.47	11.25	11.03	10.83	10.63	10.43	10.25	10.07	9.90	9.73					
6	9.567	9.411	9.259	9.113	8.971	8.834	8.700	8.571	8.446	8.324					
7	8.206	8.091	7.979	7.870	7.764	7.661	7.561	7.463	7.368	7.276					
8	7.185	7.097	7.011	6.927	6.845	6.765	6.687	6.611	6.537	6.464					
9	6.392	6.323	6.255	6.188	6.123	6.059	5.996	5.935	5.875	5.816	1'	2'	3'	4'	5'
10	5.7588	5.7023	5.6470	5.5928	5.5396	5.4874	5.4362	5.3860	5.3367	5.2883	86	172	258	344	431
11	5.2408	5.1942	5.1484	5.1034	5.0593	5.0159	4.9732	4.9313	4.8901	4.8496	72	143	215	287	359
12	4.8097	4.7706	4.7321	4.6942	4.6569	4.6202	4.5841	4.5486	4.5137	4.4793	61	121	182	242	303
13	4.4454	4.4121	4.3792	4.3469	4.3150	4.2837	4.2527	4.2223	4.1923	4.1627	52	104	156	207	260
14	4.1336	4.1048	4.0765	4.0486	4.0211	3.9939	3.9672	3.9408	3.9147	3.8890	45	90	135	180	225
15	3.8637	3.8387	3.8140	3.7897	3.7657	3.7420	3.7186	3.6955	3.6727	3.6502	39	79	118	157	196
16	3.6280	3.6060	3.5843	3.5629	3.5418	3.5209	3.5003	3.4799	3.4598	3.4399	35	69	104	138	173
17	3.4203	3.4009	3.3817	3.3628	3.3440	3.3255	3.3072	3.2891	3.2712	3.2535	31	61	92	123	153
18	3.2361	3.2188	3.2017	3.1848	3.1681	3.1515	3.1352	3.1190	3.1030	3.0872	27	55	82	110	137
19	3.0716	3.0561	3.0407	3.0256	3.0106	2.9957	2.9811	2.9665	2.9521	2.9379	25	49	74	98	123
20	2.9238	2.9099	2.8960	2.8824	2.8688	2.8555	2.8422	2.8291	2.8161	2.8032	22	44	67	89	111
21	2.7904	2.7778	2.7653	2.7529	2.7407	2.7285	2.7165	2.7046	2.6927	2.6811	20	40	60	81	101
22	2.6695	2.6580	2.6466	2.6354	2.6242	2.6131	2.6022	2.5913	2.5805	2.5699	18	37	55	73	92
23	2.5593	2.5488	2.5384	2.5282	2.5180	2.5078	2.4978	2.4879	2.4780	2.4683	17	34	50	67	84
24	2.4586	2.4490	2.4395	2.4300	2.4207	2.4114	2.4022	2.3931	2.3841	2.3751	15	31	46	62	77
25	2.3662	2.3574	2.3486	2.3400	2.3314	2.3228	2.3144	2.3060	2.2976	2.2894	14	28	43	57	71
26	2.2812	2.2730	2.2650	2.2570	2.2490	2.2412	2.2333	2.2256	2.2179	2.2103	13	26	39	52	65
27	2.2027	2.1952	2.1877	2.1803	2.1730	2.1657	2.1584	2.1513	2.1441	2.1371	12	24	36	48	61
28	2.1301	2.1231	2.1162	2.1093	2.1025	2.0957	2.0890	2.0824	2.0757	2.0692	11	22	34	45	56
29	2.0627	2.0562	2.0498	2.0434	2.0371	2.0308	2.0245	2.0183	2.0122	2.0061	10	21	31	42	52
30	2.0000	1.9940	1.9880	1.9821	1.9762	1.9703	1.9645	1.9587	1.9530	1.9473	10	19	29	39	49
31	1.9416	1.9360	1.9304	1.9249	1.9194	1.9139	1.9084	1.9031	1.8977	1.8924	9	18	27	36	45
32	1.8871	1.8818	1.8766	1.8714	1.8663	1.8612	1.8561	1.8510	1.8460	1.8410	8	17	25	34	42
33	1.8361	1.8312	1.8263	1.8214	1.8166	1.8118	1.8070	1.8023	1.7976	1.7929	8	16	24	32	40
34	1.7883	1.7837	1.7791	1.7745	1.7700	1.7655	1.7610	1.7566	1.7522	1.7478	7	15	22	30	37
35	1.7434	1.7391	1.7348	1.7305	1.7263	1.7221	1.7179	1.7137	1.7095	1.7054	7	14	21	28	35
36	1.7013	1.6972	1.6932	1.6892	1.6852	1.6812	1.6772	1.6733	1.6694	1.6655	7	13	20	26	33
37	1.6616	1.6578	1.6540	1.6502	1.6464	1.6427	1.6390	1.6353	1.6316	1.6279	6	12	19	25	31
38	1.6243	1.6207	1.6171	1.6135	1.6099	1.6064	1.6029	1.5994	1.5959	1.5925	6	12	18	23	29
39	1.5890	1.5856	1.5822	1.5788	1.5755	1.5721	1.5688	1.5655	1.5622	1.5590	6	11	17	22	28
40	1.5557	1.5525	1.5493	1.5461	1.5429	1.5398	1.5366	1.5335	1.5304	1.5273	5	10	16	21	26
41	1.5243	1.5212	1.5182	1.5151	1.5121	1.5092	1.5062	1.5032	1.5003	1.4974	5	10	15	20	25
42	1.4945	1.4916	1.4887	1.4859	1.4830	1.4802	1.4774	1.4746	1.4718	1.4690	5	9	14	19	23
43	1.4663	1.4635	1.4608	1.4581	1.4554	1.4527	1.4501	1.4474	1.4448	1.4422	4	9	13	18	22
44	1.4396	1.4370	1.4344	1.4318	1.4293	1.4267	1.4242	1.4217	1.4192	1.4167	4	8	13	17	21

$$\operatorname{cosec} A = \frac{1}{\sin A}$$

To find $\dfrac{1}{\sin 38° 23'}$, look up $\operatorname{cosec} 38° 23' = 1.6105$

Natural Cosecants

Numbers in difference columns to be *subtracted*, not added.

°	0′ 0.0°	6′ 0.1°	12′ 0.2°	18′ 0.3°	24′ 0.4°	30′ 0.5°	36′ 0.6°	42′ 0.7°	48′ 0.8°	54′ 0.9°	1′	2′	3′	4′	5′
45	1.4142	1.4118	1.4093	1.4069	1.4044	1.4020	1.3996	1.3972	1.3949	1.3925	4	8	12	16	20
46	1.3902	1.3878	1.3855	1.3832	1.3809	1.3786	1.3763	1.3741	1.3718	1.3696	4	8	11	15	19
47	1.3673	1.3651	1.3629	1.3607	1.3585	1.3563	1.3542	1.3520	1.3499	1.3478	4	7	11	14	18
48	1.3456	1.3435	1.3414	1.3393	1.3373	1.3352	1.3331	1.3311	1.3291	1.3270	3	7	10	14	17
49	1.3250	1.3230	1.3210	1.3190	1.3171	1.3151	1.3131	1.3112	1.3093	1.3073	3	7	10	13	17
50	1.3054	1.3035	1.3016	1.2997	1.2978	1.2960	1.2941	1.2923	1.2904	1.2886	3	6	9	12	16
51	1.2868	1.2849	1.2831	1.2813	1.2796	1.2778	1.2760	1.2742	1.2725	1.2708	3	6	9	12	15
52	1.2690	1.2673	1.2656	1.2639	1.2622	1.2605	1.2588	1.2571	1.2554	1.2538	3	6	8	11	14
53	1.2521	1.2505	1.2489	1.2472	1.2456	1.2440	1.2424	1.2408	1.2392	1.2376	3	5	8	11	13
54	1.2361	1.2345	1.2329	1.2314	1.2299	1.2283	1.2268	1.2253	1.2238	1.2223	3	5	8	10	13
55	1.2208	1.2193	1.2178	1.2163	1.2149	1.2134	1.2120	1.2105	1.2091	1.2076	2	5	7	10	12
56	1.2062	1.2048	1.2034	1.2020	1.2006	1.1992	1.1978	1.1964	1.1951	1.1937	2	5	7	9	12
57	1.1924	1.1910	1.1897	1.1883	1.1870	1.1857	1.1844	1.1831	1.1818	1.1805	2	4	7	9	11
58	1.1792	1.1779	1.1766	1.1753	1.1741	1.1728	1.1716	1.1703	1.1691	1.1679	2	4	6	8	10
59	1.1666	1.1654	1.1642	1.1630	1.1618	1.1606	1.1594	1.1582	1.1570	1.1559	2	4	6	8	10
60	1.1547	1.1535	1.1524	1.1512	1.1501	1.1490	1.1478	1.1467	1.1456	1.1445	2	4	6	7	9
61	1.1434	1.1423	1.1412	1.1401	1.1390	1.1379	1.1368	1.1357	1.1347	1.1336	2	4	5	7	9
62	1.1326	1.1315	1.1305	1.1294	1.1284	1.1274	1.1264	1.1253	1.1243	1.1233	2	3	5	7	9
63	1.1223	1.1213	1.1203	1.1194	1.1184	1.1174	1.1164	1.1155	1.1145	1.1136	2	3	5	6	8
64	1.1126	1.1117	1.1107	1.1098	1.1089	1.1079	1.1070	1.1061	1.1052	1.1043	2	3	5	6	8
65	1.1034	1.1025	1.1016	1.1007	1.0998	1.0989	1.0981	1.0972	1.0963	1.0955	1	3	4	6	7
66	1.0946	1.0938	1.0929	1.0921	1.0913	1.0904	1.0896	1.0888	1.0880	1.0872	1	3	4	5	7
67	1.0864	1.0856	1.0848	1.0840	1.0832	1.0824	1.0816	1.0808	1.0801	1.0793	1	3	4	5	7
68	1.0785	1.0778	1.0770	1.0763	1.0755	1.0748	1.0740	1.0733	1.0726	1.0719	1	2	4	5	6
69	1.0711	1.0704	1.0697	1.0690	1.0683	1.0676	1.0669	1.0662	1.0655	1.0649	1	2	3	5	6
70	1.0642	1.0635	1.0628	1.0622	1.0615	1.0608	1.0602	1.0595	1.0589	1.0583	1	2	3	4	5
71	1.0576	1.0570	1.0564	1.0557	1.0551	1.0545	1.0539	1.0533	1.0527	1.0521	1	2	3	4	5
72	1.0515	1.0509	1.0503	1.0497	1.0491	1.0485	1.0480	1.0474	1.0468	1.0463	1	2	3	4	5
73	1.0457	1.0451	1.0446	1.0440	1.0435	1.0429	1.0424	1.0419	1.0413	1.0408	1	2	3	4	4
74	1.0403	1.0398	1.0393	1.0388	1.0382	1.0377	1.0372	1.0367	1.0363	1.0358	1	2	2	3	4
75	1.0353	1.0348	1.0343	1.0338	1.0334	1.0329	1.0324	1.0320	1.0315	1.0311	1	2	2	3	4
76	1.0306	1.0302	1.0297	1.0293	1.0288	1.0284	1.0280	1.0276	1.0271	1.0267	1	1	2	3	4
77	1.0263	1.0259	1.0255	1.0251	1.0247	1.0243	1.0239	1.0235	1.0231	1.0227	1	1	2	3	3
78	1.0223	1.0220	1.0216	1.0212	1.0209	1.0205	1.0201	1.0198	1.0194	1.0191	1	1	2	2	3
79	1.0187	1.0184	1.0180	1.0177	1.0174	1.0170	1.0167	1.0164	1.0161	1.0157	1	1	2	2	3
80	1.0154	1.0151	1.0148	1.0145	1.0142	1.0139	1.0136	1.0133	1.0130	1.0127	0	1	1	2	2
81	1.0125	1.0122	1.0119	1.0116	1.0114	1.0111	1.0108	1.0106	1.0103	1.0101	0	1	1	2	2
82	1.0098	1.0096	1.0093	1.0091	1.0089	1.0086	1.0084	1.0082	1.0079	1.0077	0	1	1	2	2
83	1.0075	1.0073	1.0071	1.0069	1.0067	1.0065	1.0063	1.0061	1.0059	1.0057	0	1	1	1	2
84	1.0055	1.0053	1.0051	1.0050	1.0048	1.0046	1.0045	1.0043	1.0041	1.0040	0	1	1	1	1
85	1.0038	1.0037	1.0035	1.0034	1.0032	1.0031	1.0030	1.0028	1.0027	1.0026	0	0	1	1	1
86	1.0024	1.0023	1.0022	1.0021	1.0020	1.0019	1.0018	1.0017	1.0016	1.0015	0	0	0	0	1
87	1.0014	1.0013	1.0012	1.0011	1.0010	1.0010	1.0009	1.0008	1.0007	1.0007			Differences		
88	1.0006	1.0006	1.0005	1.0004	1.0004	1.0003	1.0003	1.0003	1.0002	1.0002			untrustworthy		
89	1.0002	1.0001	1.0001	1.0001	1.0001	1.0000	1.0000	1.0000	1.0000	1.0000			here		

Quadrant	Angle	cosec A =	Examples
first	0°–90°	cosec A	cosec 34° 38′ = 1.7595
second	90°–180°	cosec(180°−A)	cosec 145° 22′ = cosec(180°−145° 22′)
third	180°–270°	−cosec(A−180°)	= cosec 34° 38′ = 1.7595
fourth	270°–360°	−cosec(360°−A)	cosec 214° 38′ = −cosec(214° 38′−180°)
			= −cosec 34° 38′ = −1.7595
			cosec 325° 22′ = −cosec(360°−325° 22′)
			= −cosec 34° 38′ = −1.7595

Degrees to Radians

°	0' 0.0°	6' 0.1°	12' 0.2°	18' 0.3°	24' 0.4°	30' 0.5°	36' 0.6°	42' 0.7°	48' 0.8°	54' 0.9°	1'	2'	3'	4'	5'
0	0.0000	0.0017	0.0035	0.0052	0.0070	0.0087	0.0105	0.0122	0.0140	0.0157	3	6	9	12	15
1	0.0175	0.0192	0.0209	0.0227	0.0244	0.0262	0.0279	0.0297	0.0314	0.0332	3	6	9	12	15
2	0.0349	0.0367	0.0384	0.0401	0.0419	0.0436	0.0454	0.0471	0.0489	0.0506	3	6	9	12	15
3	0.0524	0.0541	0.0559	0.0576	0.0593	0.0611	0.0628	0.0646	0.0663	0.0681	3	6	9	12	15
4	0.0698	0.0716	0.0733	0.0750	0.0768	0.0785	0.0803	0.0820	0.0838	0.0855	3	6	9	12	15
5	0.0873	0.0890	0.0908	0.0925	0.0942	0.0960	0.0977	0.0995	0.1012	0.1030	3	6	9	12	15
6	0.1047	0.1065	0.1082	0.1100	0.1117	0.1134	0.1152	0.1169	0.1187	0.1204	3	6	9	12	15
7	0.1222	0.1239	0.1257	0.1274	0.1292	0.1309	0.1326	0.1344	0.1361	0.1379	3	6	9	12	15
8	0.1396	0.1414	0.1431	0.1449	0.1466	0.1484	0.1501	0.1518	0.1536	0.1553	3	6	9	12	15
9	0.1571	0.1588	0.1606	0.1623	0.1641	0.1658	0.1676	0.1693	0.1710	0.1728	3	6	9	12	15
10	0.1745	0.1763	0.1780	0.1798	0.1815	0.1833	0.1850	0.1868	0.1885	0.1902	3	6	9	12	15
11	0.1920	0.1937	0.1955	0.1972	0.1990	0.2007	0.2025	0.2042	0.2060	0.2077	3	6	9	12	15
12	0.2094	0.2112	0.2129	0.2147	0.2164	0.2182	0.2199	0.2217	0.2234	0.2251	3	6	9	12	15
13	0.2269	0.2286	0.2304	0.2321	0.2339	0.2356	0.2374	0.2391	0.2409	0.2426	3	6	9	12	15
14	0.2443	0.2461	0.2478	0.2496	0.2513	0.2531	0.2548	0.2566	0.2583	0.2601	3	6	9	12	15
15	0.2618	0.2635	0.2653	0.2670	0.2688	0.2705	0.2723	0.2740	0.2758	0.2775	3	6	9	12	15
16	0.2793	0.2810	0.2827	0.2845	0.2862	0.2880	0.2897	0.2915	0.2932	0.2950	3	6	9	12	15
17	0.2967	0.2985	0.3002	0.3019	0.3037	0.3054	0.3072	0.3089	0.3107	0.3124	3	6	9	12	15
18	0.3142	0.3159	0.3176	0.3194	0.3211	0.3229	0.3246	0.3264	0.3281	0.3299	3	6	9	12	15
19	0.3316	0.3334	0.3351	0.3368	0.3386	0.3403	0.3421	0.3438	0.3456	0.3473	3	6	9	12	15
20	0.3491	0.3508	0.3526	0.3543	0.3560	0.3578	0.3595	0.3613	0.3630	0.3648	3	6	9	12	15
21	0.3665	0.3683	0.3700	0.3718	0.3735	0.3752	0.3770	0.3787	0.3805	0.3822	3	6	9	12	15
22	0.3840	0.3857	0.3875	0.3892	0.3910	0.3927	0.3944	0.3962	0.3979	0.3997	3	6	9	12	15
23	0.4014	0.4032	0.4049	0.4067	0.4084	0.4102	0.4119	0.4136	0.4154	0.4171	3	6	9	12	15
24	0.4189	0.4206	0.4224	0.4241	0.4259	0.4276	0.4294	0.4311	0.4328	0.4346	3	6	9	12	15
25	0.4363	0.4381	0.4398	0.4416	0.4433	0.4451	0.4468	0.4485	0.4503	0.4520	3	6	9	12	15
26	0.4538	0.4555	0.4573	0.4590	0.4608	0.4625	0.4643	0.4660	0.4677	0.4695	3	6	9	12	15
27	0.4712	0.4730	0.4747	0.4765	0.4782	0.4800	0.4817	0.4835	0.4852	0.4869	3	6	9	12	15
28	0.4887	0.4904	0.4922	0.4939	0.4957	0.4974	0.4992	0.5009	0.5027	0.5044	3	6	9	12	15
29	0.5061	0.5079	0.5096	0.5114	0.5131	0.5149	0.5166	0.5184	0.5201	0.5219	3	6	9	12	15
30	0.5236	0.5253	0.5271	0.5288	0.5306	0.5323	0.5341	0.5358	0.5376	0.5393	3	6	9	12	15
31	0.5411	0.5428	0.5445	0.5463	0.5480	0.5498	0.5515	0.5533	0.5550	0.5568	3	6	9	12	15
32	0.5585	0.5603	0.5620	0.5637	0.5655	0.5672	0.5690	0.5707	0.5725	0.5742	3	6	9	12	15
33	0.5760	0.5777	0.5794	0.5812	0.5829	0.5847	0.5864	0.5882	0.5899	0.5917	3	6	9	12	15
34	0.5934	0.5952	0.5969	0.5986	0.6004	0.6021	0.6039	0.6056	0.6074	0.6091	3	6	9	12	15
35	0.6109	0.6126	0.6144	0.6161	0.6178	0.6196	0.6213	0.6231	0.6248	0.6266	3	6	9	12	15
36	0.6283	0.6301	0.6318	0.6336	0.6353	0.6370	0.6388	0.6405	0.6423	0.6440	3	6	9	12	15
37	0.6458	0.6475	0.6493	0.6510	0.6528	0.6545	0.6562	0.6580	0.6597	0.6615	3	6	9	12	15
38	0.6632	0.6650	0.6667	0.6685	0.6702	0.6720	0.6737	0.6754	0.6772	0.6789	3	6	9	12	15
39	0.6807	0.6824	0.6842	0.6859	0.6877	0.6894	0.6912	0.6929	0.6946	0.6964	3	6	9	12	15
40	0.6981	0.6999	0.7016	0.7034	0.7051	0.7069	0.7086	0.7103	0.7121	0.7138	3	6	9	12	15
41	0.7156	0.7173	0.7191	0.7208	0.7226	0.7243	0.7261	0.7278	0.7295	0.7313	3	6	9	12	15
42	0.7330	0.7348	0.7365	0.7383	0.7400	0.7418	0.7435	0.7453	0.7470	0.7487	3	6	9	12	15
43	0.7505	0.7522	0.7540	0.7557	0.7575	0.7592	0.7610	0.7627	0.7645	0.7662	3	6	9	12	15
44	0.7679	0.7697	0.7714	0.7732	0.7749	0.7767	0.7784	0.7802	0.7819	0.7837	3	6	9	12	15
45	0.7854	0.7871	0.7889	0.7906	0.7924	0.7941	0.7959	0.7976	0.7994	0.8011	3	6	9	12	15

$$\theta° = \frac{\pi \times \theta°}{180} = 0.01745 \times \theta \text{ radians}$$

$$30° = \frac{\pi}{6} \text{ radians} \qquad 60° = \frac{\pi}{3} \text{ radians}$$

$$45° = \frac{\pi}{4} \text{ radians} \qquad 90° = \frac{\pi}{2} \text{ radians}$$

Degrees to Radians

°	0' 0.0°	6' 0.1°	12' 0.2°	18' 0.3°	24' 0.4°	30' 0.5°	36' 0.6°	42' 0.7°	48' 0.8°	54' 0.9°	1'	2'	3'	4'	5'
45	0.7854	0.7871	0.7889	0.7906	0.7924	0.7941	0.7959	0.7976	0.7994	0.8011	3	6	9	12	15
46	0.8029	0.8046	0.8063	0.8081	0.8098	0.8116	0.8133	0.8151	0.8168	0.8186	3	6	9	12	15
47	0.8203	0.8221	0.8238	0.8255	0.8273	0.8290	0.8308	0.8325	0.8343	0.8360	3	6	9	12	15
48	0.8378	0.8395	0.8412	0.8430	0.8447	0.8465	0.8482	0.8500	0.8517	0.8535	3	6	9	12	15
49	0.8552	0.8570	0.8587	0.8604	0.8622	0.8639	0.8657	0.8674	0.8692	0.8709	3	6	9	12	15
50	0.8727	0.8744	0.8762	0.8779	0.8796	0.8814	0.8831	0.8849	0.8866	0.8884	3	6	9	12	15
51	0.8901	0.8919	0.8936	0.8954	0.8971	0.8988	0.9006	0.9023	0.9041	0.9058	3	6	9	12	15
52	0.9076	0.9093	0.9111	0.9128	0.9146	0.9163	0.9180	0.9198	0.9215	0.9233	3	6	9	12	15
53	0.9250	0.9268	0.9285	0.9303	0.9320	0.9338	0.9355	0.9372	0.9390	0.9407	3	6	9	12	15
54	0.9425	0.9442	0.9460	0.9477	0.9495	0.9512	0.9529	0.9547	0.9564	0.9582	3	6	9	12	15
55	0.9599	0.9617	0.9634	0.9652	0.9669	0.9687	0.9704	0.9721	0.9739	0.9756	3	6	9	12	15
56	0.9774	0.9791	0.9809	0.9826	0.9844	0.9861	0.9879	0.9896	0.9913	0.9931	3	6	9	12	15
57	0.9948	0.9966	0.9983	1.0001	1.0018	1.0036	1.0053	1.0071	1.0088	1.0105	3	6	9	12	15
58	1.0123	1.0140	1.0158	1.0175	1.0193	1.0210	1.0228	1.0245	1.0263	1.0280	3	6	9	12	15
59	1.0297	1.0315	1.0332	1.0350	1.0367	1.0385	1.0402	1.0420	1.0437	1.0455	3	6	9	12	15
60	1.0472	1.0489	1.0507	1.0524	1.0542	1.0559	1.0577	1.0594	1.0612	1.0629	3	6	9	12	15
61	1.0647	1.0664	1.0681	1.0699	1.0716	1.0734	1.0751	1.0769	1.0786	1.0804	3	6	9	12	15
62	1.0821	1.0838	1.0856	1.0873	1.0891	1.0908	1.0926	1.0943	1.0961	1.0978	3	6	9	12	15
63	1.0996	1.1013	1.1030	1.1048	1.1065	1.1083	1.1100	1.1118	1.1135	1.1153	3	6	9	12	15
64	1.1170	1.1188	1.1205	1.1222	1.1240	1.1257	1.1275	1.1292	1.1310	1.1327	3	6	9	12	15
65	1.1345	1.1362	1.1380	1.1397	1.1414	1.1432	1.1449	1.1467	1.1484	1.1502	3	6	9	12	15
66	1.1519	1.1537	1.1554	1.1572	1.1589	1.1606	1.1624	1.1641	1.1659	1.1676	3	6	9	12	15
67	1.1694	1.1711	1.1729	1.1746	1.1764	1.1781	1.1798	1.1816	1.1833	1.1851	3	6	9	12	15
68	1.1868	1.1886	1.1903	1.1921	1.1938	1.1956	1.1973	1.1990	1.2008	1.2025	3	6	9	12	15
69	1.2043	1.2060	1.2078	1.2095	1.2113	1.2130	1.2147	1.2165	1.2182	1.2200	3	6	9	12	15
70	1.2217	1.2235	1.2252	1.2270	1.2287	1.2305	1.2322	1.2339	1.2357	1.2374	3	6	9	12	15
71	1.2392	1.2409	1.2427	1.2444	1.2462	1.2479	1.2497	1.2514	1.2531	1.2549	3	6	9	12	15
72	1.2566	1.2584	1.2601	1.2619	1.2636	1.2654	1.2671	1.2689	1.2706	1.2723	3	6	9	12	15
73	1.2741	1.2758	1.2776	1.2793	1.2811	1.2828	1.2846	1.2863	1.2881	1.2898	3	6	9	12	15
74	1.2915	1.2933	1.2950	1.2968	1.2985	1.3003	1.3020	1.3038	1.3055	1.3073	3	6	9	12	15
75	1.3090	1.3107	1.3125	1.3142	1.3160	1.3177	1.3195	1.3212	1.3230	1.3247	3	6	9	12	15
76	1.3265	1.3282	1.3299	1.3317	1.3334	1.3352	1.3369	1.3387	1.3404	1.3422	3	6	9	12	15
77	1.3439	1.3456	1.3474	1.3491	1.3509	1.3526	1.3544	1.3561	1.3579	1.3596	3	6	9	12	15
78	1.3614	1.3631	1.3648	1.3666	1.3683	1.3701	1.3718	1.3736	1.3753	1.3771	3	6	9	12	15
79	1.3788	1.3806	1.3823	1.3840	1.3858	1.3875	1.3893	1.3910	1.3928	1.3945	3	6	9	12	15
80	1.3963	1.3980	1.3998	1.4015	1.4032	1.4050	1.4067	1.4085	1.4102	1.4120	3	6	9	12	15
81	1.4137	1.4155	1.4172	1.4190	1.4207	1.4224	1.4242	1.4259	1.4277	1.4294	3	6	9	12	15
82	1.4312	1.4329	1.4347	1.4364	1.4382	1.4399	1.4416	1.4434	1.4451	1.4469	3	6	9	12	15
83	1.4486	1.4504	1.4521	1.4539	1.4556	1.4573	1.4591	1.4608	1.4626	1.4643	3	6	9	12	15
84	1.4661	1.4678	1.4696	1.4713	1.4731	1.4748	1.4765	1.4783	1.4800	1.4818	3	6	9	12	15
85	1.4835	1.4853	1.4870	1.4888	1.4905	1.4923	1.4940	1.4957	1.4975	1.4992	3	6	9	12	15
86	1.5010	1.5027	1.5045	1.5062	1.5080	1.5097	1.5115	1.5132	1.5149	1.5167	3	6	9	12	15
87	1.5184	1.5202	1.5219	1.5237	1.5254	1.5272	1.5289	1.5307	1.5324	1.5341	3	6	9	12	15
88	1.5359	1.5376	1.5394	1.5411	1.5429	1.5446	1.5464	1.5481	1.5499	1.5516	3	6	9	12	15
89	1.5533	1.5551	1.5568	1.5586	1.5603	1.5621	1.5638	1.5656	1.5673	1.5691	3	6	9	12	15

85° 46' = 1.4969 radians (direct from table)

0.6219 radians = 35° 38' (by finding 0.6219 in the table and reading off the corresponding angle in degrees)

Table of Squares

x	0	1	2	3	4	5	6	7	8	9	1	2	3	4	5	6	7	8	9
1·0	1·000	1·020	1·040	1·061	1·082	1·103	1·124	1·145	1·166	1·188	2	4	6	8	10	13	15	17	19
1·1	1·210	1·232	1·254	1·277	1·300	1·323	1·346	1·369	1·392	1·416	2	5	7	9	11	14	16	18	21
1·2	1·440	1·464	1·488	1·513	1·538	1·563	1·588	1·613	1·638	1·664	2	5	7	10	12	15	17	20	22
1·3	1·690	1·716	1·742	1·769	1·796	1·823	1·850	1·877	1·904	1·932	3	5	8	11	13	16	19	22	24
1·4	1·960	1·988	2·016	2·045	2·074	2·103	2·132	2·161	2·190	2·220	3	6	9	12	14	17	20	23	26
1·5	2·250	2·280	2·310	2·341	2·372	2·403	2·434	2·465	2·496	2·528	3	6	9	12	15	19	22	25	28
1·6	2·560	2·592	2·624	2·657	2·690	2·723	2·756	2·789	2·822	2·856	3	7	10	13	16	20	23	26	30
1·7	2·890	2·924	2·958	2·993	3·028	3·063	3·098	3·133	3·168	3·204	3	7	10	14	17	21	24	28	31
1·8	3·240	3·276	3·312	3·349	3·386	3·423	3·460	3·497	3·534	3·572	4	7	11	15	18	22	26	30	33
1·9	3·610	3·648	3·686	3·725	3·764	3·803	3·842	3·881	3·920	3·960	4	8	12	16	19	23	27	31	35
2·0	4·000	4·040	4·080	4·121	4·162	4·203	4·244	4·285	4·326	4·368	4	8	12	16	20	25	29	33	37
2·1	4·410	4·452	4·494	4·537	4·580	4·623	4·666	4·709	4·752	4·796	4	9	13	17	21	26	30	34	39
2·2	4·840	4·884	4·928	4·973	5·018	5·063	5·108	5·153	5·198	5·244	4	9	13	18	22	27	31	36	40
2·3	5·290	5·336	5·382	5·429	5·476	5·523	5·570	5·617	5·664	5·712	5	9	14	19	23	28	33	38	42
2·4	5·760	5·808	5·856	5·905	5·954	6·003	6·052	6·101	6·150	6·200	5	10	15	20	24	29	34	39	44
2·5	6·250	6·300	6·350	6·401	6·452	6·503	6·554	6·605	6·656	6·708	5	10	15	20	25	31	36	41	46
2·6	6·760	6·812	6·864	6·917	6·970	7·023	7·076	7·129	7·182	7·236	5	11	16	21	26	32	37	42	48
2·7	7·290	7·344	7·398	7·453	7·508	7·563	7·618	7·673	7·728	7·784	5	11	16	22	27	33	38	44	49
2·8	7·840	7·896	7·952	8·009	8·066	8·123	8·180	8·237	8·294	8·352	6	11	17	23	28	34	40	46	51
2·9	8·410	8·468	8·526	8·585	8·644	8·703	8·762	8·821	8·880	8·940	6	12	18	24	29	35	41	47	53
3·0	9·000	9·060	9·120	9·181	9·242	9·303	9·364	9·425	9·486	9·548	6	12	18	24	30	37	43	49	55
3·1	9·610	9·672	9·734	9·797	9·860	9·923	9·986	10·05	10·11	10·18	6	13	19	25	31	38	44	50	57
3·2	10·24	10·30	10·37	10·43	10·50	10·56	10·63	10·69	10·76	10·82	1	1	2	3	3	4	5	5	6
3·3	10·89	10·96	11·02	11·09	11·16	11·22	11·29	11·36	11·42	11·49	1	1	2	3	3	4	5	5	6
3·4	11·56	11·63	11·70	11·76	11·83	11·90	11·97	12·04	12·11	12·18	1	1	2	3	3	4	5	6	6
3·5	12·25	12·32	12·39	12·46	12·53	12·60	12·67	12·74	12·82	12·89	1	1	2	3	4	4	5	6	6
3·6	12·96	13·03	13·10	13·18	13·25	13·32	13·40	13·47	13·54	13·62	1	1	2	3	4	4	5	6	7
3·7	13·69	13·76	13·84	13·91	13·99	14·06	14·14	14·21	14·29	14·36	1	2	2	3	4	4	5	6	7
3·8	14·44	14·52	14·59	14·67	14·75	14·82	14·90	14·98	15·05	15·13	1	2	2	3	4	5	5	6	7
3·9	15·21	15·29	15·37	15·44	15·52	15·60	15·68	15·76	15·84	15·92	1	2	2	3	4	5	6	6	7
4·0	16·00	16·08	16·16	16·24	16·32	16·40	16·48	16·56	16·65	16·73	1	2	2	3	4	5	6	6	7
4·1	16·81	16·89	16·97	17·06	17·14	17·22	17·31	17·39	17·47	17·56	1	2	2	3	4	5	6	7	7
4·2	17·64	17·72	17·81	17·89	17·98	18·06	18·15	18·23	18·32	18·40	1	2	3	3	4	5	6	7	8
4·3	18·49	18·58	18·66	18·75	18·84	18·92	19·01	19·10	19·18	19·27	1	2	3	3	4	5	6	7	8
4·4	19·36	19·45	19·54	19·62	19·71	19·80	19·89	19·98	20·07	20·16	1	2	3	4	4	5	6	7	8
4·5	20·25	20·34	20·43	20·52	20·61	20·70	20·79	20·88	20·98	21·07	1	2	3	4	5	5	6	7	8
4·6	21·16	21·25	21·34	21·44	21·53	21·62	21·72	21·81	21·90	22·00	1	2	3	4	5	6	7	7	8
4·7	22·09	22·18	22·28	22·37	22·47	22·56	22·66	22·75	22·85	22·94	1	2	3	4	5	6	7	8	9
4·8	23·04	23·14	23·23	23·33	23·43	23·52	23·62	23·72	23·81	23·91	1	2	3	4	5	6	7	8	9
4·9	24·01	24·11	24·21	24·30	24·40	24·50	24·60	24·70	24·80	24·90	1	2	3	4	5	6	7	8	9
5·0	25·00	25·10	25·20	25·30	25·40	25·50	25·60	25·70	25·81	25·91	1	2	3	4	5	6	7	8	9
5·1	26·01	26·11	26·21	26·32	26·42	26·52	26·63	26·73	26·83	26·94	1	2	3	4	5	6	7	8	9
5·2	27·04	27·14	27·25	27·35	27·46	27·56	27·67	27·77	27·88	27·98	1	2	3	4	5	6	7	8	9
5·3	28·09	28·20	28·30	28·41	28·52	28·62	28·73	28·84	28·94	29·05	1	2	3	4	5	6	7	9	10
5·4	29·16	29·27	29·38	29·48	29·59	29·70	29·81	29·92	30·03	30·14	1	2	3	4	5	7	8	9	10

The table of squares give the squares of numbers from 1 to 10. To find the squares of numbers greater than 10 proceed as follows:

To find $(468.8)^2$

$$(468.8)^2 = (4.688 \times 100)^2 = 4.688^2 \times 100^2 = 21.97 \times 10^4$$
$$\text{or } 219\,700$$

Table of Squares

x	0	1	2	3	4	5	6	7	8	9	1	2	3	4	5	6	7	8	9
5·5	30·25	30·36	30·47	30·58	30·69	30·80	30·91	31·02	31·14	31·25	1	2	3	4	6	7	8	9	10
5·6	31·36	31·47	31·58	31·70	31·81	31·92	32·04	32·15	32·26	32·38	1	2	3	5	6	7	8	9	10
5·7	32·49	32·60	32·72	32·83	32·95	33·06	33·18	33·29	33·41	33·52	1	2	3	5	6	7	8	9	10
5·8	33·64	33·76	33·87	33·99	34·11	34·22	34·34	34·46	34·57	34·69	1	2	4	5	6	7	8	9	11
5·9	34·81	34·93	35·05	35·16	35·28	35·40	35·52	35·64	35·76	35·88	1	2	4	5	6	7	8	10	11
6·0	36·00	36·12	36·24	36·36	36·48	36·60	36·72	36·84	36·97	37·09	1	2	4	5	6	7	9	10	11
6·1	37·21	37·33	37·45	37·58	37·70	37·82	37·95	38·07	38·19	38·32	1	2	4	5	6	7	9	10	11
6·2	38·44	38·56	38·69	38·81	38·94	39·06	39·19	39·31	39·44	39·56	1	3	4	5	6	8	9	10	11
6·3	39·69	39·82	39·94	40·07	40·20	40·32	40·45	40·58	40·70	40·83	1	3	4	5	6	8	9	10	11
6·4	40·96	41·09	41·22	41·34	41·47	41·60	41·73	41·86	41·99	42·12	1	3	4	5	6	8	9	10	12
6·5	42·25	42·38	42·51	42·64	42·77	42·90	43·03	43·16	43·30	43·43	1	3	4	5	7	8	9	10	12
6·6	43·56	43·69	43·82	43·96	44·09	44·22	44·36	44·49	44·62	44·76	1	3	4	5	7	8	9	11	12
6·7	44·89	45·02	45·16	45·29	45·43	45·56	45·70	45·83	45·97	46·10	1	3	4	5	7	8	9	11	12
6·8	46·24	46·38	46·51	56·65	46·79	46·92	47·06	47·20	47·33	47·47	1	3	4	5	7	8	10	11	12
6·9	47·61	47·75	47·89	48·02	48·16	48·30	48·44	48·58	48·72	48·86	1	3	4	6	7	8	10	11	13
7·0	49·00	49·14	49·28	49·42	49·56	49·70	49·84	49·98	50·13	50·27	1	3	4	6	7	8	10	11	13
7·1	50·41	50·55	50·69	50·84	50·98	51·12	51·27	51·41	51·55	51·70	1	3	4	6	7	9	10	11	13
7·2	51·84	51·98	52·13	52·27	52·42	52·56	52·71	52·85	53·00	53·14	1	3	4	6	7	9	10	12	13
7·3	53·29	53·44	53·58	53·73	53·88	54·02	54·17	54·32	54·46	54·61	1	3	4	6	7	9	10	12	13
7·4	54·76	54·91	55·06	55·20	55·35	55·50	55·65	55·80	55·95	56·10	1	3	4	6	7	9	10	12	13
7·5	56·25	56·40	56·55	56·70	56·85	57·00	57·15	57·30	57·46	57·61	2	3	5	6	8	9	11	12	14
7·6	57·76	57·91	58·06	58·22	58·37	58·52	58·68	58·83	58·98	59·14	2	3	5	6	8	9	11	12	14
7·7	59·29	59·44	59·60	59·75	59·91	60·06	60·22	60·37	60·53	60·68	2	3	5	6	8	9	11	12	14
7·8	60·84	61·00	61·15	61·31	61·47	61·62	61·78	61·94	62·09	62·25	2	3	5	6	8	9	11	13	14
7·9	62·41	62·57	62·73	62·88	63·04	63·20	63·36	63·52	63·68	63·84	2	3	5	6	8	10	11	13	14
8·0	64·00	64·16	64·32	64·48	64·64	64·80	64·96	65·12	65·29	65·45	2	3	5	6	8	10	11	13	14
8·1	65·61	65·77	65·93	66·10	66·26	66·42	67·59	66·75	66·91	67·08	2	3	5	7	8	10	11	13	15
8·2	67·24	67·40	67·57	67·73	67·90	68·06	68·23	68·39	68·56	68·72	2	3	5	7	8	10	12	13	15
8·3	68·89	69·06	69·22	69·39	69·56	69·72	69·89	70·06	70·22	70·39	2	3	5	7	8	10	12	13	15
8·4	70·56	70·73	70·90	71·06	71·23	71·40	71·57	71·74	71·91	72·08	2	3	5	7	8	10	12	14	15
8·5	72·25	72·42	72·59	72·76	72·93	73·10	73·27	73·44	73·62	73·79	2	3	5	7	9	10	12	14	15
8·6	73·96	74·13	74·30	74·48	74·65	74·82	75·00	75·17	75·34	75·52	2	3	5	7	9	10	12	14	16
8·7	75·69	75·86	76·04	76·21	76·39	76·56	76·74	76·91	77·09	77·26	2	4	5	7	9	11	12	14	16
8·8	77·44	77·62	77·79	77·97	78·15	78·32	78·50	78·68	78·85	79·03	2	4	5	7	9	11	12	14	16
8·9	79·21	79·39	79·57	79·74	79·92	80·10	80·28	80·46	80·64	80·82	2	4	5	7	9	11	13	14	16
9·0	81·00	81·18	81·36	81·54	81·72	81·90	82·08	82·26	82·45	82·63	2	4	5	7	9	11	13	14	16
9·1	82·81	82·99	83·17	83·36	83·54	83·72	83·91	84·09	84·27	84·46	2	4	5	7	9	11	13	15	16
9·2	84·64	84·82	85·01	85·19	85·38	85·56	85·75	85·93	86·12	86·30	2	4	6	7	9	11	13	15	17
9·3	86·49	86·68	86·86	87·05	87·24	87·42	87·61	87·80	87·98	88·17	2	4	6	7	9	11	13	15	17
9·4	88·36	88·55	88·74	88·92	89·11	89·30	89·49	89·68	89·87	90·06	2	4	6	8	9	11	13	15	17
9·5	90·25	90·44	90·63	90·82	91·01	91·20	91·39	91·58	91·78	91·97	2	4	6	8	10	11	13	15	17
9·6	92·16	92·35	92·54	92·74	92·93	93·12	93·32	93·51	93·70	93·90	2	4	6	8	10	12	14	15	17
9·7	94·09	94·28	94·48	94·67	94·87	95·06	95·26	95·45	95·65	95·84	2	4	6	8	10	12	14	16	18
9·8	96·04	96·24	96·43	96·63	96·83	97·02	97·22	97·42	97·61	97·81	2	4	6	8	10	12	14	16	18
9·9	98·01	98·21	98·41	98·60	98·80	99·00	99·20	99·40	99·60	99·80	2	4	6	8	10	12	14	16	18

To find $(0.2388)^2$

$$(0.2388)^2 = (2.388 \times \tfrac{1}{10})^2 = (2.388)^2 \times (\tfrac{1}{10})^2$$
$$= 5.702 \times \tfrac{1}{100} = 5.702 \times 10^{-2} \text{ or } 0.05\,702$$

Table of Square Roots from 1–10

	·0	·1	·2	·3	·4	·5	·6	·7	·8	·9	1	2	3	4	5	6	7	8	9
1·0	1·000	1·005	1·010	1·015	1·020	1·025	1·030	1·034	1·039	1·044	0	1	1	2	2	3	3	4	4
1·1	1·049	1·054	1·058	1·063	1·068	1·072	1·077	1·082	1·086	1·091	0	1	1	2	2	3	3	4	4
1·2	1·095	1·100	1·105	1·109	1·114	1·118	1·122	1·127	1·131	1·136	0	1	1	2	2	3	3	4	4
1·3	1·140	1·145	1·149	1·153	1·158	1·162	1·166	1·170	1·175	1·179	0	1	1	2	2	3	3	3	4
1·4	1·183	1·187	1·192	1·196	1·200	1·204	1·208	1·212	1·217	1·221	0	1	1	2	2	3	3	3	4
1·5	1·225	1·229	1·233	1·237	1·241	1·245	1·249	1·253	1·257	1·261	0	1	1	2	2	2	3	3	4
1·6	1·265	1·269	1·273	1·277	1·281	1·285	1·288	1·292	1·296	1·300	0	1	1	2	2	2	3	3	3
1·7	1·304	1·308	1·311	1·315	1·319	1·323	1·327	1·330	1·334	1·338	0	1	1	1	2	2	3	3	3
1·8	1·342	1·345	1·349	1·353	1·356	1·360	1·364	1·367	1·371	1·375	0	1	1	1	2	2	3	3	3
1·9	1·378	1·382	1·386	1·389	1·393	1·396	1·400	1·404	1·407	1·411	0	1	1	1	2	2	3	3	3
2·0	1·414	1·418	1·421	1·425	1·428	1·432	1·435	1·439	1·442	1·446	0	1	1	1	2	2	2	3	3
2·1	1·449	1·453	1·456	1·459	1·463	1·466	1·470	1·473	1·476	1·480	0	1	1	1	2	2	2	3	3
2·2	1·483	1·487	1·490	1·493	1·497	1·500	1·503	1·507	1·510	1·513	0	1	1	1	2	2	2	3	3
2·3	1·517	1·520	1·523	1·526	1·530	1·533	1·536	1·539	1·543	1·546	0	1	1	1	2	2	2	2	3
2·4	1·549	1·552	1·556	1·559	1·562	1·565	1·568	1·572	1·575	1·578	0	1	1	1	2	2	2	2	3
2·5	1·581	1·584	1·587	1·591	1·594	1·597	1·600	1·603	1·606	1·609	0	1	1	1	2	2	2	2	3
2·6	1·612	1·616	1·619	1·622	1·625	1·628	1·631	1·634	1·637	1·640	0	1	1	1	2	2	2	2	3
2·7	1·643	1·646	1·649	1·652	1·655	1·658	1·661	1·664	1·667	1·670	0	1	1	1	2	2	2	2	3
2·8	1·673	1·676	1·679	1·682	1·685	1·688	1·691	1·694	1·697	1·700	0	1	1	1	2	2	2	2	3
2·9	1·703	1·706	1·709	1·712	1·715	1·718	1·720	1·723	1·726	1·729	0	1	1	1	1	2	2	2	3
3·0	1·732	1·735	1·738	1·741	1·744	1·746	1·749	1·752	1·755	1·758	0	1	1	1	1	2	2	2	3
3·1	1·761	1·764	1·766	1·769	1·772	1·775	1·778	1·780	1·783	1·786	0	1	1	1	1	2	2	2	3
3·2	1·789	1·792	1·794	1·797	1·800	1·803	1·806	1·808	1·811	1·814	0	1	1	1	1	2	2	2	2
3·3	1·817	1·819	1·822	1·825	1·828	1·830	1·833	1·836	1·838	1·841	0	1	1	1	1	2	2	2	2
3·4	1·844	1·847	1·849	1·852	1·855	1·857	1·860	1·863	1·865	1·868	0	1	1	1	1	2	2	2	2
3·5	1·871	1·873	1·876	1·879	1·881	1·884	1·887	1·889	1·892	1·895	0	1	1	1	1	2	2	2	2
3·6	1·897	1·900	1·903	1·905	1·908	1·910	1·913	1·916	1·918	1·921	0	1	1	1	1	2	2	2	2
3·7	1·924	1·926	1·929	1·931	1·934	1·936	1·939	1·942	1·944	1·947	0	1	1	1	1	2	2	2	2
3·8	1·949	1·952	1·954	1·957	1·960	1·962	1·965	1·967	1·970	1·972	0	1	1	1	1	2	2	2	2
3·9	1·975	1·977	1·980	1·982	1·985	1·987	1·990	1·992	1·995	1·997	0	0	1	1	1	1	2	2	2
4·0	2·000	2·002	2·005	2·007	2·010	2·012	2·015	2·017	2·020	2·022	0	0	1	1	1	1	2	2	2
4·1	2·025	2·027	2·030	2·032	2·035	2·037	2·040	2·042	2·045	2·047	0	0	1	1	1	1	2	2	2
4·2	2·049	2·052	2·054	2·057	2·059	2·062	2·064	2·066	2·069	2·071	0	0	1	1	1	1	2	2	2
4·3	2·074	2·076	2·078	2·081	2·083	2·086	2·088	2·090	2·093	2·095	0	0	1	1	1	1	1	2	2
4·4	2·098	2·100	2·102	2·105	2·107	2·110	2·112	2·114	2·117	2·119	0	0	1	1	1	1	1	2	2
4·5	2·121	2·124	2·126	2·128	2·131	2·133	2·135	2·138	2·140	2·142	0	0	1	1	1	1	1	2	2
4·6	2·145	2·147	2·149	2·152	2·154	2·156	2·159	2·161	2·163	2·166	0	0	1	1	1	1	1	2	2
4·7	2·168	2·170	2·173	2·175	2·177	2·179	2·182	2·184	2·186	2·189	0	0	1	1	1	1	1	2	2
4·8	2·191	2·193	2·195	2·198	2·200	2·202	2·205	2·207	2·209	2·211	0	0	1	1	1	1	1	2	2
4·9	2·214	2·216	2·218	2·220	2·223	2·225	2·227	2·229	2·232	2·234	0	0	1	1	1	1	1	2	2
5·0	2·236	2·238	2·241	2·243	2·245	2·247	2·249	2·252	2·254	2·256	0	0	1	1	1	1	1	2	2
5·1	2·258	2·261	2·263	2·265	2·267	2·269	2·272	2·274	2·276	2·278	0	0	1	1	1	1	1	2	2
5·2	2·280	2·283	2·285	2·287	2·289	2·291	2·293	2·296	2·298	2·300	0	0	1	1	1	1	1	2	2
5·3	2·302	2·304	2·307	2·309	2·311	2·313	2·315	2·317	2·319	2·322	0	0	1	1	1	1	1	2	2
5·4	2·324	2·326	2·328	2·330	2·332	2·335	2·337	2·339	2·341	2·343	0	0	1	1	1	1	1	2	2

There are two sets of square root tables one giving square roots of numbers between 1 and 10 and the other, square roots of numbers between 10 and 100.

$$\sqrt{1.873} = 1.368 \text{ (using square roots of numbers between 1 and 10)}$$
$$\sqrt{18.73} = 4.327 \text{ (using square roots of numbers between 10 and 100)}$$

Table of Square Roots from 1–10

	·0	·1	·2	·3	·4	·5	·6	·7	·8	·9	1	2	3	4	5	6	7	8	9
5·5	2·345	2·347	2·349	2·352	2·354	2·356	2·358	2·360	2·362	2·364	0	0	1	1	1	1	1	2	2
5·6	2·366	2·369	2·371	2·373	2·375	2·377	2·379	2·381	2·383	2·385	0	0	1	1	1	1	1	2	2
5·7	2·387	2·390	2·392	2·394	2·396	2·398	2·400	2·402	2·404	2·406	0	0	1	1	1	1	1	2	2
5·8	2·408	2·410	2·412	2·415	2·417	2·419	2·421	2·423	2·425	2·427	0	0	1	1	1	1	1	2	2
5·9	2·429	2·431	2·433	2·435	2·437	2·439	2·441	2·443	2·445	2·447	0	0	1	1	1	1	1	2	2
6·0	2·449	2·452	2·454	2·456	2·458	2·460	2·462	2·464	2·466	2·468	0	0	1	1	1	1	1	2	2
6·1	2·470	2·472	2·474	2·476	2·478	2·480	2·482	2·484	2·486	2·488	0	0	1	1	1	1	1	2	2
6·2	2·490	2·492	2·494	2·496	2·498	2·500	2·502	2·504	2·506	2·508	0	0	1	1	1	1	1	2	2
6·3	2·510	2·512	2·514	2·516	2·518	2·520	2·522	2·524	2·526	2·528	0	0	1	1	1	1	1	2	2
6·4	2·530	2·532	2·534	2·536	2·538	2·540	2·542	2·544	2·546	2·548	0	0	1	1	1	1	1	2	2
6·5	2·550	2·551	2·553	2·555	2·557	2·559	2·561	2·563	2·565	2·567	0	0	1	1	1	1	1	2	2
6·6	2·569	2·571	2·573	2·575	2·577	2·579	2·581	2·583	2·585	2·587	0	0	1	1	1	1	1	2	2
6·7	2·588	2·590	2·592	2·594	2·596	2·598	2·600	2·602	2·604	2·606	0	1	1	1	1	1	1	2	2
6·8	2·608	2·610	2·612	2·613	2·615	2·617	2·619	2·621	2·623	2·625	0	0	1	1	1	1	1	2	2
6·9	2·627	2·629	2·631	2·632	2·634	2·636	2·638	2·640	2·642	2·644	0	0	1	1	1	1	1	2	2
7·0	2·646	2·648	2·650	2·651	2·653	2·655	2·657	2·659	2·661	2·663	0	0	1	1	1	1	1	2	2
7·1	2·665	2·666	2·668	2·670	2·672	2·674	2·676	2·678	2·680	2·681	0	0	1	1	1	1	1	2	2
7·2	2·683	2·685	2·687	2·689	2·691	2·693	2·694	2·696	2·698	2·700	0	0	1	1	1	1	1	2	2
7·3	2·702	2·704	2·706	2·707	2·709	2·711	2·713	2·715	2·717	2·718	0	0	1	1	1	1	1	2	2
7·4	2·720	2·722	2·724	2·726	2·728	2·729	2·731	2·733	2·735	2·737	0	0	1	1	1	1	1	2	2
7·5	2·739	2·740	2·742	2·744	2·746	2·748	2·750	2·751	2·753	2·755	0	0	1	1	1	1	1	2	2
7·6	2·757	2·759	2·760	2·762	2·764	2·766	2·768	2·769	2·771	2·773	0	0	1	1	1	1	1	2	2
7·7	2·775	2·777	2·778	2·780	2·782	2·784	2·786	2·787	2·789	2·791	0	0	1	1	1	1	1	2	2
7·8	2·793	2·795	2·796	2·798	2·800	2·802	2·804	2·805	2·807	2·809	0	0	1	1	1	1	1	2	2
7·9	2·811	2·812	2·814	2·816	2·818	2·820	2·821	2·823	2·825	2·827	0	0	1	1	1	1	1	2	2
8·0	2·828	2·830	2·832	2·834	2·835	2·837	2·839	2·841	2·843	2·844	0	0	1	1	1	1	1	2	2
8·1	2·846	2·848	2·850	2·851	2·853	2·855	2·857	2·858	2·860	2·862	0	0	1	1	1	1	1	2	2
8·2	2·864	2·865	2·867	2·869	2·871	2·872	2·874	2·876	2·877	2·879	0	0	1	1	1	1	1	2	2
8·3	2·881	2·883	2·884	2·886	2·888	2·890	2·891	2·893	2·895	2·897	0	0	1	1	1	1	1	2	2
8·4	2·898	2·900	2·902	2·903	2·905	2·907	2·909	2·910	2·912	2·914	0	0	1	1	1	1	1	2	2
8·5	2·915	2·917	2·919	2·921	2·922	2·924	2·926	2·927	2·929	2·931	0	0	1	1	1	1	1	2	2
8·6	2·933	2·934	2·936	2·938	2·939	2·941	2·943	2·944	2·946	2·948	0	0	1	1	1	1	1	2	2
8·7	2·950	2·951	2·953	2·955	2·956	2·958	2·960	2·961	2·963	2·965	0	0	1	1	1	1	1	2	2
8·8	2·966	2·968	2·970	2·972	2·973	2·975	2·977	2·978	2·980	2·982	0	0	1	1	1	1	1	2	2
8·9	2·983	2·985	2·987	2·988	2·990	2·992	2·993	2·995	2·997	2·998	0	0	1	1	1	1	1	2	2
9·0	3·000	3·002	3·003	3·005	3·007	3·008	3·010	3·012	3·013	3·015	0	0	0	1	1	1	1	1	1
9·1	3·017	3·018	3·020	3·022	3·023	3·025	3·027	3·028	3·030	3·032	0	0	0	1	1	1	1	1	1
9·2	3·033	3·035	3·036	3·038	3·040	3·041	3·043	3·045	3·046	3·048	0	0	0	1	1	1	1	1	1
9·3	3·050	3·051	3·053	3·055	3·056	3·058	3·059	3·061	3·063	3·064	0	0	0	1	1	1	1	1	1
9·4	3·066	3·068	3·069	3·071	3·072	3·074	3·076	3·077	3·079	3·081	0	0	0	1	1	1	1	1	1
9·5	3·082	3·084	3·085	3·087	3·089	3·090	3·092	3·094	3·095	3·097	0	0	0	1	1	1	1	1	1
9·6	3·098	3·100	3·102	3·103	3·105	3·106	3·108	3·110	3·111	3·113	0	0	0	1	1	1	1	1	1
9·7	3·114	3·116	3·118	3·119	3·121	3·122	3·124	3·126	3·127	3·129	0	0	0	1	1	1	1	1	1
9·8	3·130	3·132	3·134	3·135	3·137	3·138	3·140	3·142	3·143	3·145	0	0	0	1	1	1	1	1	1
9·9	3·146	3·148	3·150	3·151	3·153	3·154	3·156	3·158	3·159	3·161	0	0	0	1	1	1	1	1	1

To find $\sqrt{836.3}$

Mark off figures in pairs to the left of the decimal point. Thus 836.3 becomes 8'36.3. The first period is 8 so look up $\sqrt{8.363} = 2.892$. For each period to the left of the decimal there will be one figure to the left of the decimal point in the answer. Thus $\sqrt{836.3} = 28.92$.

To find $\sqrt{8363}$.

Marking off in pairs 8363 becomes 83'63 so look up $\sqrt{83.63} = 9.145$ and hence $\sqrt{8363} = 91.45$.

Table of Square Roots from 10–100

x	·0	·1	·2	·3	·4	·5	·6	·7	·8	·9	1	2	3	4	5	6	7	8	9
10	3·162	3·178	3·194	3·209	3·225	3·240	3·256	3·271	3·286	3·302	2	3	5	6	8	10	11	13	14
11	3·317	3·332	3·347	3·362	3·376	3·391	3·406	3·421	3·435	3·450	1	3	4	6	7	9	10	12	13
12	3·464	3·479	3·493	3·507	3·521	3·536	3·550	3·564	3·578	3·592	1	3	4	6	7	8	10	11	13
13	3·606	3·619	3·633	3·647	3·661	3·674	3·688	3·701	3·715	3·728	1	3	4	5	7	8	10	11	12
14	3·742	3·755	3·768	3·782	3·795	3·808	3·821	3·834	3·847	3·860	1	3	4	5	7	8	9	10	12
15	3·873	3·886	3·899	3·912	3·924	3·937	3·950	3·962	3·975	3·987	1	3	4	5	6	8	9	10	11
16	4·000	4·012	4·025	4·037	4·050	4·062	4·074	4·087	4·099	4·111	1	2	4	5	6	7	9	10	11
17	4·123	4·135	4·147	4·159	4·171	4·183	4·195	4·207	4·219	4·231	1	2	4	5	6	7	8	10	11
18	4·243	4·254	4·266	4·278	4·290	4·301	4·313	4·324	4·336	4·347	1	2	3	5	6	7	8	9	10
19	4·359	4·370	4·382	4·393	4·405	4·416	4·427	4·438	4·450	4·461	1	2	3	5	6	7	8	9	10
20	4·472	4·483	4·494	4·506	4·517	4·528	4·539	4·550	4·561	4·572	1	2	3	4	6	7	8	9	10
21	4·583	4·593	4·604	4·615	4·626	4·637	4·648	4·658	4·669	4·680	1	2	3	4	5	6	8	9	10
22	4·690	4·701	4·712	4·722	4·733	4·743	4·754	4·764	4·775	4·785	1	2	3	4	5	6	7	8	10
23	4·796	4·806	4·817	4·827	4·837	4·848	4·858	4·868	4·879	4·889	1	2	3	4	5	6	7	8	9
24	4·899	4·909	4·919	4·930	4·940	4·950	4·960	4·970	4·980	4·990	1	2	3	4	5	6	7	8	9
25	5·000	5·010	5·020	5·030	5·040	5·050	5·060	5·070	5·079	5·089	1	2	3	4	5	6	7	8	9
26	5·099	5·109	5·119	5·128	5·138	5·148	5·158	5·167	5·177	5·187	1	2	3	4	5	6	7	8	9
27	5·196	5·206	5·215	5·225	5·235	5·244	5·254	5·263	5·273	5·282	1	2	3	4	5	6	7	8	9
28	5·292	5·301	5·310	5·320	5·329	5·339	5·348	5·357	5·367	5·376	1	2	3	4	5	6	7	7	8
29	5·385	5·394	5·404	5·413	5·422	5·431	5·441	5·450	5·459	5·468	1	2	3	4	5	5	6	7	8
30	5·477	5·486	5·495	5·505	5·514	5·523	5·532	5·541	5·550	5·559	1	2	3	4	5	5	6	7	8
31	5·568	5·577	5·586	5·595	5·604	5·612	5·621	5·630	5·639	5·648	1	2	3	4	4	5	6	7	8
32	5·657	5·666	5·675	5·683	5·692	5·701	5·710	5·718	5·727	5·736	1	2	3	4	4	5	6	7	8
33	5·745	5·753	5·762	5·771	5·779	5·788	5·797	5·805	5·814	5·822	1	2	3	3	4	5	6	7	8
34	5·831	5·840	5·848	5·857	5·865	5·874	5·882	5·891	5·899	5·908	1	2	3	3	4	5	6	7	8
35	5·916	5·925	5·933	5·941	5·950	5·958	5·967	5·975	5·983	5·992	1	2	2	3	4	5	6	7	8
36	6·000	6·008	6·017	6·025	6·033	6·042	6·050	6·058	6·066	6·075	1	2	2	3	4	5	6	7	7
37	6·083	6·091	6·099	6·107	6·116	6·124	6·132	6·140	6·148	6·156	1	2	2	3	4	5	6	6	7
38	6·164	6·173	6·181	6·189	6·197	6·205	6·213	6·221	6·229	6·237	1	2	2	3	4	5	6	6	7
39	6·245	6·253	6·261	6·269	6·277	6·285	6·293	6·301	6·309	6·317	1	2	2	3	4	5	6	6	7
40	6·325	6·332	6·340	6·348	6·356	6·364	6·372	6·380	6·387	6·395	1	2	2	3	4	5	6	6	7
41	6·403	6·411	6·419	6·427	6·434	6·442	6·450	6·458	6·465	6·473	1	2	2	3	4	5	5	6	7
42	6·481	6·488	6·496	6·504	6·512	6·519	6·527	6·535	6·542	6·550	1	2	2	3	4	5	5	6	7
43	6·557	6·565	6·573	6·580	6·588	6·595	6·603	6·611	6·618	6·626	1	2	2	3	4	5	5	6	7
44	6·633	6·641	6·648	6·656	6·663	6·671	6·678	6·686	6·693	6·701	1	2	2	3	4	4	5	6	7
45	6·708	6·716	6·723	6·731	6·738	6·745	6·753	6·760	6·768	6·775	1	1	2	3	4	4	5	6	7
46	6·782	6·790	6·797	6·804	6·812	6·819	6·826	6·834	6·841	6·848	1	1	2	3	4	4	5	6	7
47	6·856	6·863	6·870	6·877	6·885	6·892	6·899	6·907	6·914	6·921	1	1	2	3	4	4	5	6	6
48	6·928	6·935	6·943	6·950	6·957	6·964	6·971	6·979	6·986	6·993	1	1	2	3	4	4	5	6	6
49	7·000	7·007	7·014	7·021	7·029	7·036	7·043	7·050	7·057	7·064	1	1	2	3	4	4	5	6	6
50	7·071	7·078	7·085	7·092	7·099	7·106	7·113	7·120	7·127	7·134	1	1	2	3	4	4	5	6	6
51	7·141	7·148	7·155	7·162	7·169	7·176	7·183	7·190	7·197	7·204	1	1	2	3	4	4	5	6	6
52	7·211	7·218	7·225	7·232	7·239	7·246	7·253	7·259	7·266	7·273	1	1	2	3	3	4	5	5	6
53	7·280	7·287	7·294	7·301	7·308	7·314	7·321	7·328	7·335	7·342	1	1	2	3	3	4	5	5	6
54	7·348	7·355	7·362	7·369	7·376	7·382	7·389	7·396	7·403	7·409	1	1	2	3	3	4	5	5	6

To find $\sqrt{0.000\,2168}$

In the case of decimal numbers mark off in pairs to the right of the decimal point. Thus 0.02 168 becomes 0.00′02′16′8. Apart from the zero pair the first pair is 02 so look up $\sqrt{2.168} = 1.473$. For each zero pair in the original number there will be one zero following the decimal point in the answer. Therefore $\sqrt{0.000\,2168} = 0.014\,73$.

Table of Square Roots from 10–100

x	·0	·1	·2	·3	·4	·5	·6	·7	·8	·9	1	2	3	4	5	6	7	8	9
55	7·416	7·423	7·430	7·436	7·443	7·450	7·457	7·463	7·470	7·477	1	1	2	3	3	4	5	5	6
56	7·483	7·490	7·497	7·503	7·510	7·517	7·523	7·530	7·537	7·543	1	1	2	3	3	4	5	5	6
57	7·550	7·556	7·563	7·570	7·576	7·583	7·589	7·596	7·603	7·609	1	1	2	3	3	4	5	5	6
58	7·616	7·622	7·629	7·635	7·642	7·649	7·655	7·662	7·668	7·675	1	1	2	3	3	4	5	5	6
59	7·681	7·688	7·694	7·701	7·707	7·714	7·720	7·727	7·733	7·740	1	1	2	3	3	4	5	5	6
60	7·746	7·752	7·759	7·765	7·772	7·778	7·785	7·791	7·797	7·804	1	1	2	3	3	4	4	5	6
61	7·810	7·817	7·823	7·829	7·836	7·842	7·849	7·855	7·861	7·868	1	1	2	2	3	4	4	5	5
62	7·874	7·880	7·887	7·893	7·899	7·906	7·912	7·918	7·925	7·931	1	1	2	2	3	4	4	5	5
63	7·937	7·944	7·950	7·956	7·962	7·969	7·975	7·981	7·987	7·994	1	1	2	2	3	4	4	5	5
64	8·000	8·006	8·012	8·019	8·025	8·031	8·037	8·044	8·050	8·056	1	1	2	2	3	4	4	5	5
65	8·062	8·068	8·075	8·081	8·087	8·093	8·099	8·106	8·112	8·118	1	1	2	2	3	4	4	5	5
66	8·124	8·130	8·136	8·142	8·149	8·155	8·161	8·167	8·173	8·179	1	1	2	2	3	4	4	5	5
67	8·185	8·191	8·198	8·204	8·210	8·216	8·222	8·228	8·234	8 240	1	1	2	2	3	4	4	5	5
68	8·246	8·252	8·258	8·264	8·270	8·276	8·283	8·289	8·295	8·301	·1	1	2	2	3	4	4	5	5
69	8·307	8·313	8·319	8·325	8·331	8·337	8·343	8·349	8·355	8·361	1	1	2	2	3	4	4	5	5
70	8·367	8·373	8·379	8·385	8·390	8·396	8·402	8·408	8·414	8·420	1	1	2	2	3	4	4	5	5
71	8·426	6·432	8·438	8·444	8·450	8·456	8·462	8·468	8·473	8·479	1	1	2	2	3	3	4	5	5
72	8·485	8·491	8·497	8·503	8·509	8·515	8·521	8·526	8·532	8·538	1	1	2	2	3	3	4	5	5
73	8·544	8·550	8·556	8·562	8·567	8·573	8·579	8·585	8·591	8 597	1	1	2	2	3	3	4	5	5
74	8·602	8·608	8·614	8·620	8·626	8·631	8·637	8·643	8·649	8·654	1	1	2	2	3	3	4	5	5
75	8·660	8·666	8·672	8·678	8·683	8·689	8·695	8·701	8·706	8·712	1	1	2	2	3	3	4	4	5
76	8·718	8·724	8·729	8·735	8·741	8·746	8·752	8·758	8·764	8·769	1	1	2	2	3	3	4	4	5
77	8·775	8·781	8·786	8·792	8·798	8·803	8·809	8·815	8·820	8·826	1	1	2	2	3	3	4	4	5
78	8·832	8·837	8·843	8·849	8·854	8·860	8·866	8·871	8·877	8·883	1	1	2	2	3	3	4	4	5
79	8·888	8·894	8·899	8·905	8·911	8·916	8·922	8·927	8·933	8·939	1	1	2	2	3	3	4	4	5
80	8·944	8·950	8·955	8·961	8·967	8·972	8·978	8·983	8·989	8·994	1	1	2	2	3	3	4	4	5
81	9·000	9·006	9·011	9·017	9·022	9·028	9·033	9·039	9·044	9·050	1	1	2	2	3	3	4	4	5
82	9·055	9·061	9·066	9·072	9·077	9·083	9·088	9·094	9·099	9·105	1	1	2	2	3	3	4	4	5
83	9·110	9·116	9·121	9·127	9·132	9·138	9·143	9·149	9·154	9·160	1	1	2	2	3	3	4	4	5
84	9·165	9·171	9·176	9·182	9·187	9·192	9·198	9·203	9·209	9·214	1	1	2	2	3	3	4	4	5
85	9·220	9·225	9·230	9·236	9·241	9·247	9·252	9·257	9·263	9·268	1	1	2	2	3	3	4	4	5
86	9·274	9·279	9·284	9·290	9·295	9·301	9·306	9·311	9·317	9·322	1	1	2	2	3	3	4	4	5
87	9·327	9·333	9·338	9·343	9·349	9·354	9·359	9·365	9·370	9·375	1	1	2	2	3	3	4	4	5
88	9·381	9·386	9·391	9·397	9·402	9·407	9·413	9·418	9·423	9·429	1	1	2	2	3	3	4	4	5
89	9·434	9·439	9·445	9·450	9·455	9·460	9·466	9·471	9·476	9·482	1	1	2	2	3	3	4	4	5
90	9·487	9·492	9·497	9·503	9·508	9·513	9·518	9·524	9·529	9·534	1	1	2	2	3	3	4	4	5
91	9·539	9·545	9·550	9·555	9·560	9·566	9·571	9·576	9·581	9·586	1	1	2	2	3	3	4	4	5
92	9·592	9·597	9·602	9·607	9·612	9·618	9·623	9·628	9·633	9·638	1	1	2	2	3	3	4	4	5
93	9·644	9·649	9·654	9·659	9·664	9·670	9·675	9·680	9·685	9·690	1	1	2	2	3	3	4	4	5
94	9·695	9·701	9·706	9·711	9·716	9·721	9·726	9·731	9·737	9·742	1	1	2	2	3	3	4	4	5
95	9·747	9·752	9·757	9·762	9·767	9·772	9·778	9·783	9·788	9·793	1	1	2	2	3	3	4	4	5
96	9·798	9·803	9·808	9·813	9·818	9·823	9·829	9·834	9·839	9·844	1	1	2	2	3	3	4	4	5
97	9·849	9·854	9·859	9·864	9·869	9·874	9·879	9·884	9·889	9·894	1	1	2	2	3	3	4	4	5
98	9·899	9·905	9·910	9·915	9·920	9·925	9·930	9·935	9·940	9·945	1	1	2	2	3	3	4	4	5
99	9·950	9·955	9·960	9·965	9·970	9·975	9·980	9·985	9·990	9·995	0	1	1	2	2	3	4	4	4

To find $\sqrt{0.002\,168}$

Marking off in pairs 0.002 168 becomes 0.00′21′68 so look up $\sqrt{21.68} = 4.657$

and hence $\sqrt{0.002\,168} = 0.046\,57$.

Table of Reciprocals of Numbers from 1–10

Numbers in difference columns to be *subtracted*, not added

	0	1	2	3	4	5	6	7	8	9	1	2	3	4	5	6	7	8	9
1.0	1.0000	0.9901	0.9804	0.9709	0.9615	0.9524	0.9434	0.9346	0.9259	0.9174									
1.1	0.9091	0.9009	0.8929	0.8850	0.8772	0.8696	0.8621	0.8547	0.8475	0.8403									
1.2	0.8333	0.8264	0.8197	0.8130	0.8065	0.8000	0.7937	0.7874	0.7813	0.7752									
1.3	0.7692	0.7634	0.7576	0.7519	0.7463	0.7407	0.7353	0.7299	0.7246	0.7194									
1.4	0.7143	0.7092	0.7042	0.6993	0.6944	0.6897	0.6849	0.6803	0.6757	0.6711									
1.5	0.6667	0.6623	0.6579	0.6536	0.6494	0.6452	0.6410	0.6369	0.6329	0.6289	4	8	12	17	21	25	29	33	37
1.6	0.6250	0.6211	0.6173	0.6135	0.6098	0.6061	0.6024	0.5988	0.5952	0.5917	4	7	11	15	18	22	26	29	33
1.7	0.5882	0.5848	0.5814	0.5780	0.5747	0.5714	0.5682	0.5650	0.5618	0.5587	3	7	10	13	16	20	23	26	29
1.8	0.5556	0.5525	0.5495	0.5464	0.5435	0.5405	0.5376	0.5348	0.5319	0.5291	3	6	9	12	15	18	20	23	26
1.9	0.5263	0.5236	0.5208	0.5181	0.5155	0.5128	0.5102	0.5076	0.5051	0.5025	3	5	8	11	13	16	18	21	24
2.0	0.5000	0.4975	0.4950	0.4926	0.4902	0.4878	0.4854	0.4831	0.4808	0.4785	2	5	7	10	12	14	17	19	21
2.1	0.4762	0.4739	0.4717	0.4695	0.4673	0.4651	0.4630	0.4608	0.4587	0.4566	2	4	6	9	11	13	15	17	19
2.2	0.4545	0.4525	0.4505	0.4484	0.4464	0.4444	0.4425	0.4405	0.4386	0.4367	2	4	6	8	10	12	14	16	18
2.3	0.4348	0.4329	0.4310	0.4292	0.4274	0.4255	0.4237	0.4219	0.4202	0.4184	2	4	5	7	9	11	13	14	16
2.4	0.4167	0.4149	0.4132	0.4115	0.4098	0.4082	0.4065	0.4049	0.4032	0.4016	2	3	5	7	8	10	12	13	15
2.5	0.4000	0.3984	0.3968	0.3953	0.3937	0.3922	0.3906	0.3891	0.3876	0.3861	2	3	5	6	8	9	11	12	14
2.6	0.3846	0.3831	0.3817	0.3802	0.3788	0.3774	0.3759	0.3745	0.3731	0.3717	1	3	4	6	7	9	10	11	13
2.7	0.3704	0.3690	0.3676	0.3663	0.3650	0.3636	0.3623	0.3610	0.3597	0.3584	1	3	4	5	7	8	9	11	12
2.8	0.3571	0.3559	0.3546	0.3534	0.3521	0.3509	0.3497	0.3484	0.3472	0.3460	1	2	4	5	6	7	9	10	11
2.9	0.3448	0.3436	0.3425	0.3413	0.3401	0.3390	0.3378	0.3367	0.3356	0.3344	1	2	3	5	6	7	8	9	10
3.0	0.3333	0.3322	0.3311	0.3300	0.3289	0.3279	0.3268	0.3257	0.3247	0.3236	1	2	3	4	5	6	8	9	10
3.1	0.3226	0.3215	0.3205	0.3195	0.3185	0.3175	0.3165	0.3155	0.3145	0.3135	1	2	3	4	5	6	7	8	9
3.2	0.3125	0.3115	0.3106	0.3096	0.3086	0.3077	0.3067	0.3058	0.3049	0.3040	1	2	3	4	5	6	7	8	9
3.3	0.3030	0.3021	0.3012	0.3003	0.2994	0.2985	0.2976	0.2967	0.2959	0.2950	1	2	3	4	4	5	6	7	8
3.4	0.2941	0.2933	0.2924	0.2915	0.2907	0.2899	0.2890	0.2882	0.2874	0.2865	1	2	3	3	4	5	6	7	8
3.5	0.2857	0.2849	0.2841	0.2833	0.2825	0.2817	0.2809	0.2801	0.2793	0.2786	1	2	2	3	4	5	6	6	7
3.6	0.2778	0.2770	0.2762	0.2755	0.2747	0.2740	0.2732	0.2725	0.2717	0.2710	1	2	2	3	4	5	5	6	7
3.7	0.2703	0.2695	0.2688	0.2681	0.2674	0.2667	0.2660	0.2653	0.2646	0.2639	1	1	2	3	4	4	5	6	6
3.8	0.2632	0.2625	0.2618	0.2611	0.2604	0.2597	0.2591	0.2584	0.2577	0.2571	1	1	2	3	3	4	5	5	6
3.9	0.2564	0.2558	0.2551	0.2545	0.2538	0.2532	0.2525	0.2519	0.2513	0.2506	1	1	2	3	3	4	4	5	6
4.0	0.2500	0.2494	0.2488	0.2481	0.2475	0.2469	0.2463	0.2457	0.2451	0.2445	1	1	2	2	3	4	4	5	5
4.1	0.2439	0.2433	0.2427	0.2421	0.2415	0.2410	0.2404	0.2398	0.2392	0.2387	1	1	2	2	3	3	4	5	5
4.2	0.2381	0.2375	0.2370	0.2364	0.2358	0.2353	0.2347	0.2342	0.2336	0.2331	1	1	2	2	3	3	4	4	5
4.3	0.2326	0.2320	0.2315	0.2309	0.2304	0.2299	0.2294	0.2288	0.2283	0.2278	1	1	2	2	3	3	4	4	5
4.4	0.2273	0.2268	0.2262	0.2257	0.2252	0.2247	0.2242	0.2237	0.2232	0.2227	1	1	2	2	3	3	4	4	5
4.5	0.2222	0.2217	0.2212	0.2208	0.2203	0.2198	0.2193	0.2188	0.2183	0.2179	0	1	1	2	2	3	3	4	4
4.6	0.2174	0.2169	0.2165	0.2160	0.2155	0.2151	0.2146	0.2141	0.2137	0.2132	0	1	1	2	2	3	3	4	4
4.7	0.2128	0.2123	0.2119	0.2114	0.2110	0.2105	0.2101	0.2096	0.2092	0.2088	0	1	1	2	2	3	3	4	4
4.8	0.2083	0.2079	0.2075	0.2070	0.2066	0.2062	0.2058	0.2053	0.2049	0.2045	0	1	1	2	2	3	3	3	4
4.9	0.2041	0.2037	0.2033	0.2028	0.2024	0.2020	0.2016	0.2012	0.2008	0.2004	0	1	1	2	2	2	3	3	4
5.0	0.2000	0.1996	0.1992	0.1988	0.1984	0.1980	0.1976	0.1972	0.1969	0.1965	0	1	1	2	2	2	3	3	4
5.1	0.1961	0.1957	0.1953	0.1949	0.1946	0.1942	0.1938	0.1934	0.1931	0.1927	0	1	1	2	2	2	3	3	3
5.2	0.1923	0.1919	0.1916	0.1912	0.1908	0.1905	0.1901	0.1898	0.1894	0.1890	0	1	1	1	2	2	3	3	3
5.3	0.1887	0.1883	0.1880	0.1876	0.1873	0.1869	0.1866	0.1862	0.1859	0.1855	0	1	1	1	2	2	2	3	3
5.4	0.1852	0.1848	0.1845	0.1842	0.1838	0.1835	0.1832	0.1828	0.1825	0.1821	0	1	1	1	2	2	2	3	3

Reciprocal of a number $= \dfrac{1}{\text{number}}$

Reciprocal of 2.361 $= \dfrac{1}{2.361} = 0.4235$

To find reciprocals of numbers outside the range of 1 to 10 proceed as shown on page 51.

Table of Reciprocals of Numbers from 1–10

Numbers in difference columns to be *subtracted*, not added.

	0	1	2	3	4	5	6	7	8	9	1 2 3	4 5 6	7 8 9
5.5	0.1818	0.1815	0.1812	0.1808	0.1805	0.1802	0.1799	0.1795	0.1792	0.1789	0 1 1	1 2 2	2 3 3
5.6	0.1786	0.1783	0.1779	0.1776	0.1773	0.1770	0.1767	0.1764	0.1761	0.1757	0 1 1	1 2 2	2 3 3
5.7	0.1754	0.1751	0.1748	0.1745	0.1742	0.1739	0.1736	0.1733	0.1730	0.1727	0 1 1	1 2 2	2 2 3
5.8	0.1724	0.1721	0.1718	0.1715	0.1712	0.1709	0.1706	0.1704	0.1701	0.1698	0 1 1	1 1 2	2 2 3
5.9	0.1695	0.1692	0.1689	0.1686	0.1684	0.1681	0.1678	0.1675	0.1672	0.1669	0 1 1	1 1 2	2 2 3
6.0	0.1667	0.1664	0.1661	0.1658	0.1656	0.1653	0.1650	0.1647	0.1645	0.1642	0 1 1	1 1 2	2 2 2
6.1	0.1639	0.1637	0.1634	0.1631	0.1629	0.1626	0.1623	0.1621	0.1618	0.1616	0 1 1	1 1 2	2 2 2
6.2	0.1613	0.1610	0.1608	0.1605	0.1603	0.1600	0.1597	0.1595	0.1592	0.1590	0 1 1	1 1 2	2 2 2
6.3	0.1587	0.1585	0.1582	0.1580	0.1577	0.1575	0.1572	0.1570	0.1567	0.1565	0 0 1	1 1 1	2 2 2
6.4	0.1563	0.1560	0.1558	0.1555	0.1553	0.1550	0.1548	0.1546	0.1543	0.1541	0 0 1	1 1 1	2 2 2
6.5	0.1538	0.1536	0.1534	0.1531	0.1529	0.1527	0.1524	0.1522	0.1520	0.1517	0 0 1	1 1 1	2 2 2
6.6	0.1515	0.1513	0.1511	0.1508	0.1506	0.1504	0.1502	0.1499	0.1497	0.1495	0 0 1	1 1 1	2 2 2
6.7	0.1493	0.1490	0.1488	0.1486	0.1484	0.1481	0.1479	0.1477	0.1475	0.1473	0 0 1	1 1 1	2 2 2
6.8	0.1471	0.1468	0.1466	0.1464	0.1462	0.1460	0.1458	0.1456	0.1453	0.1451	0 0 1	1 1 1	1 2 2
6.9	0.1449	0.1447	0.1445	0.1443	0.1441	0.1439	0.1437	0.1435	0.1433	0.1431	0 0 1	1 1 1	1 2 2
7.0	0.1429	0.1427	0.1425	0.1422	0.1420	0.1418	0.1416	0.1414	0.1412	0.1410	0 0 1	1 1 1	1 2 2
7.1	0.1408	0.1406	0.1404	0.1403	0.1401	0.1399	0.1397	0.1395	0.1393	0.1391	0 0 1	1 1 1	1 2 2
7.2	0.1389	0.1387	0.1385	0.1383	0.1381	0.1379	0.1377	0.1376	0.1374	0.1372	0 0 1	1 1 1	1 2 2
7.3	0.1370	0.1368	0.1366	0.1364	0.1362	0.1361	0.1359	0.1357	0.1355	0.1353	0 0 1	1 1 1	1 1 2
7.4	0.1351	0.1350	0.1348	0.1346	0.1344	0.1342	0.1340	0.1339	0.1337	0.1335	0 0 1	1 1 1	1 1 2
7.5	0.1333	0.1332	0.1330	0.1328	0.1326	0.1325	0.1323	0.1321	0.1319	0.1318	0 0 1	1 1 1	1 1 2
7.6	0.1316	0.1314	0.1312	0.1311	0.1309	0.1307	0.1305	0.1304	0.1302	0.1300	0 0 1	1 1 1	1 1 2
7.7	0.1299	0.1297	0.1295	0.1294	0.1292	0.1290	0.1289	0.1287	0.1285	0.1284	0 0 0	1 1 1	1 1 1
7.8	0.1282	0.1280	0.1279	0.1277	0.1276	0.1274	0.1272	0.1271	0.1269	0.1267	0 0 0	1 1 1	1 1 1
7.9	0.1266	0.1264	0.1263	0.1261	0.1259	0.1258	0.1256	0.1255	0.1253	0.1252	0 0 0	1 1 1	1 1 1
8.0	0.1250	0.1248	0.1247	0.1245	0.1244	0.1242	0.1241	0.1239	0.1238	0.1236	0 0 0	1 1 1	1 1 1
8.1	0.1235	0.1233	0.1232	0.1230	0.1229	0.1227	0.1225	0.1224	0.1222	0.1221	0 0 0	1 1 1	1 1 1
8.2	0.1220	0.1218	0.1217	0.1215	0.1214	0.1212	0.1211	0.1209	0.1208	0.1206	0 0 0	1 1 1	1 1 1
8.3	0.1205	0.1203	0.1202	0.1200	0.1199	0.1198	0.1196	0.1195	0.1193	0.1192	0 0 0	1 1 1	1 1 1
8.4	0.1190	0.1189	0.1188	0.1186	0.1185	0.1183	0.1182	0.1181	0.1179	0.1178	0 0 0	1 1 1	1 1 1
8.5	0.1176	0.1175	0.1174	0.1172	0.1171	0.1170	0.1168	0.1167	0.1166	0.1164	0 0 0	1 1 1	1 1 1
8.6	0.1163	0.1161	0.1160	0.1159	0.1157	0.1156	0.1155	0.1153	0.1152	0.1151	0 0 0	1 1 1	1 1 1
8.7	0.1149	0.1148	0.1147	0.1145	0.1144	0.1143	0.1142	0.1140	0.1139	0.1138	0 0 0	1 1 1	1 1 1
8.8	0.1136	0.1135	0.1134	0.1133	0.1131	0.1130	0.1129	0.1127	0.1126	0.1125	0 0 0	1 1 1	1 1 1
8.9	0.1124	0.1122	0.1121	0.1120	0.1119	0.1117	0.1116	0.1115	0.1114	0.1112	0 0 0	0 1 1	1 1 1
9.0	0.1111	0.1110	0.1109	0.1107	0.1106	0.1105	0.1104	0.1103	0.1101	0.1100	0 0 0	0 1 1	1 1 1
9.1	0.1099	0.1098	0.1096	0.1095	0.1094	0.1093	0.1092	0.1091	0.1089	0.1088	0 0 0	0 1 1	1 1 1
9.2	0.1087	0.1086	0.1085	0.1083	0.1082	0.1081	0.1080	0.1079	0.1078	0.1076	0 0 0	0 1 1	1 1 1
9.3	0.1075	0.1074	0.1073	0.1072	0.1071	0.1070	0.1068	0.1067	0.1066	0.1065	0 0 0	0 1 1	1 1 1
9.4	0.1064	0.1063	0.1062	0.1060	0.1059	0.1058	0.1057	0.1056	0.1055	0.1054	0 0 0	0 1 1	1 1 1
9.5	0.1053	0.1052	0.1050	0.1049	0.1048	0.1047	0.1046	0.1045	0.1044	0.1043	0 0 0	0 1 1	1 1 1
9.6	0.1042	0.1041	0.1040	0.1038	0.1037	0.1036	0.1035	0.1034	0.1033	0.1032	0 0 0	0 1 1	1 1 1
9.7	0.1031	0.1030	0.1029	0.1028	0.1027	0.1026	0.1025	0.1024	0.1022	0.1021	0 0 0	0 1 1	1 1 1
9.8	0.1020	0.1019	0.1018	0.1017	0.1016	0.1015	0.1014	0.1013	0.1012	0.1011	0 0 0	0 1 1	1 1 1
9.9	0.1010	0.1009	0.1008	0.1007	0.1006	0.1005	0.1004	0.1003	0.1002	0.1001	0 0 0	0 1 1	1 1 1

To find $\dfrac{1}{639.2}$

$$\frac{1}{639.2} = \frac{1}{6.392} \times \frac{1}{100} = 0.1565 \times \frac{1}{100} = 0.1565 \times 10^{-2}$$

or 0.001 565

To find $\dfrac{1}{0.039\,82}$

$$\frac{1}{0.039\,82} = \frac{1}{3.982} \times \frac{100}{1} = 0.2512 \times 10^{2} = 25.12$$

Natural Logarithms

	0	1	2	3	4	5	6	7	8	9	1	2	3	4	5	6	7	8	9
1.0	0.0000	0.0100	0.0198	0.0296	0.0392	0.0488	0.0583	0.0677	0.0770	0.0862	10	19	29	38	48	57	67	76	86
1.1	0.0953	0.1044	0.1133	0.1222	0.1310	0.1398	0.1484	0.1570	0.1655	0.1740	9	17	26	35	44	52	61	70	78
1.2	0.1823	0.1906	0.1989	0.2070	0.2151	0.2231	0.2311	0.2390	0.2469	0.2546	8	16	24	32	40	48	56	64	72
1.3	0.2624	0.2700	0.2776	0.2852	0.2927	0.3001	0.3075	0.3148	0.3221	0.3293	7	15	22	30	37	44	52	59	67
1.4	0.3365	0.3436	0.3507	0.3577	0.3646	0.3716	0.3784	0.3853	0.3920	0.3988	7	14	21	28	34	41	48	55	62
1.5	0.4055	0.4121	0.4187	0.4253	0.4318	0.4383	0.4447	0.4511	0.4574	0.4637	6	13	19	26	32	39	45	52	58
1.6	0.4700	0.4762	0.4824	0.4886	0.4947	0.5008	0.5068	0.5128	0.5188	0.5247	6	12	18	24	30	36	42	48	55
1.7	0.5306	0.5365	0.5423	0.5481	0.5539	0.5596	0.5653	0.5710	0.5766	0.5822	6	11	17	23	29	34	40	46	51
1.8	0.5878	0.5933	0.5988	0.6043	0.6098	0.6152	0.6206	0.6259	0.6313	0.6366	5	11	16	22	27	32	38	43	49
1.9	0.6419	0.6471	0.6523	0.6575	0.6627	0.6678	0.6729	0.6780	0.6831	0.6881	5	10	15	21	26	31	36	41	46
2.0	0.6931	0.6981	0.7031	0.7080	0.7129	0.7178	0.7227	0.7275	0.7324	0.7372	5	10	15	20	24	29	34	39	44
2.1	0.7419	0.7467	0.7514	0.7561	0.7608	0.7655	0.7701	0.7747	0.7793	0.7839	5	9	14	19	23	28	33	37	42
2.2	0.7885	0.7930	0.7975	0.8020	0.8065	0.8109	0.8154	0.8198	0.8242	0.8286	4	9	13	18	22	27	31	36	40
2.3	0.8329	0.8372	0.8416	0.8459	0.8502	0.8544	0.8587	0.8629	0.8671	0.8713	4	9	13	17	21	26	30	34	38
2.4	0.8755	0.8796	0.8838	0.8879	0.8920	0.8961	0.9002	0.9042	0.9083	0.9123	4	8	12	16	20	24	29	33	37
2.5	0.9163	0.9203	0.9243	0.9282	0.9322	0.9361	0.9400	0.9439	0.9478	0.9517	4	8	12	16	20	24	27	31	35
2.6	0.9555	0.9594	0.9632	0.9670	0.9708	0.9746	0.9783	0.9821	0.9858	0.9895	4	8	11	15	19	23	26	30	34
2.7	0.9933	0.9969	1.0006	1.0043	1.0080	1.0116	1.0152	1.0188	1.0225	1.0260	4	7	11	15	18	22	25	29	33
2.8	1.0296	1.0332	1.0367	1.0403	1.0438	1.0473	1.0508	1.0543	1.0578	1.0613	4	7	11	14	18	21	25	28	32
2.9	1.0647	1.0682	1.0716	1.0750	1.0784	1.0818	1.0852	1.0886	1.0919	1.0953	3	7	10	14	17	20	24	27	31
3.0	1.0986	1.1019	1.1053	1.1086	1.1119	1.1151	1.1184	1.1217	1.1249	1.1282	3	7	10	13	16	20	23	26	30
3.1	1.1314	1.1346	1.1378	1.1410	1.1442	1.1474	1.1506	1.1537	1.1569	1.1600	3	6	10	13	16	19	22	25	29
3.2	1.1632	1.1663	1.1694	1.1725	1.1756	1.1787	1.1817	1.1848	1.1878	1.1909	3	6	9	12	15	18	22	25	28
3.3	1.1939	1.1969	1.2000	1.2030	1.2060	1.2090	1.2119	1.2149	1.2179	1.2208	3	6	9	12	15	18	21	24	27
3.4	1.2238	1.2267	1.2296	1.2326	1.2355	1.2384	1.2413	1.2442	1.2470	1.2499	3	6	9	12	14	17	20	23	26
3.5	1.2528	1.2556	1.2585	1.2613	1.2641	1.2669	1.2698	1.2726	1.2754	1.2782	3	6	8	11	14	17	20	23	25
3.6	1.2809	1.2837	1.2865	1.2892	1.2920	1.2947	1.2975	1.3002	1.3029	1.3056	3	5	8	11	14	16	19	22	25
3.7	1.3083	1.3110	1.3137	1.3164	1.3191	1.3218	1.3244	1.3271	1.3297	1.3324	3	5	8	11	13	16	19	21	24
3.8	1.3350	1.3376	1.3403	1.3429	1.3455	1.3481	1.3507	1.3533	1.3558	1.3584	3	5	8	10	13	16	18	21	23
3.9	1.3610	1.3635	1.3661	1.3686	1.3712	1.3737	1.3762	1.3788	1.3813	1.3838	3	5	8	10	13	15	18	20	23
4.0	1.3863	1.3888	1.3913	1.3938	1.3962	1.3987	1.4012	1.4036	1.4061	1.4085	2	5	7	10	12	15	17	20	22
4.1	1.4110	1.4134	1.4159	1.4183	1.4207	1.4231	1.4255	1.4279	1.4303	1.4327	2	5	7	10	12	14	17	19	22
4.2	1.4351	1.4375	1.4398	1.4422	1.4446	1.4469	1.4493	1.4516	1.4540	1.4563	2	5	7	9	12	14	16	19	21
4.3	1.4586	1.4609	1.4633	1.4656	1.4679	1.4702	1.4725	1.4748	1.4770	1.4793	2	5	7	9	11	14	16	18	21
4.4	1.4816	1.4839	1.4861	1.4884	1.4907	1.4929	1.4951	1.4974	1.4996	1.5019	2	4	7	9	11	13	16	18	20
4.5	1.5041	1.5063	1.5085	1.5107	1.5129	1.5151	1.5173	1.5195	1.5217	1.5239	2	4	7	9	11	13	15	18	20
4.6	1.5261	1.5282	1.5304	1.5326	1.5347	1.5369	1.5390	1.5412	1.5433	1.5454	2	4	6	9	11	13	15	17	19
4.7	1.5476	1.5497	1.5518	1.5539	1.5560	1.5581	1.5602	1.5623	1.5644	1.5665	2	4	6	8	11	13	15	17	19
4.8	1.5686	1.5707	1.5728	1.5748	1.5769	1.5790	1.5810	1.5831	1.5851	1.5877	2	4	6	8	10	12	14	16	19
4.9	1.5892	1.5913	1.5933	1.5953	1.5974	1.5994	1.6014	1.6034	1.6054	1.6074	2	4	6	8	10	12	14	16	18
5.0	1.6094	1.6114	1.6134	1.6154	1.6174	1.6194	1.6214	1.6233	1.6253	1.6273	2	4	6	8	10	12	14	16	18
5.1	1.6292	1.6312	1.6332	1.6351	1.6371	1.6390	1.6409	1.6429	1.6448	1.6467	2	4	6	8	10	12	14	16	18
5.2	1.6487	1.6506	1.6525	1.6544	1.6563	1.6582	1.6601	1.6620	1.6639	1.6658	2	4	6	8	10	11	13	15	17
5.3	1.6677	1.6696	1.6715	1.6734	1.6752	1.6771	1.6790	1.6808	1.6827	1.6845	2	4	6	7	9	11	13	15	17
5.4	1.6864	1.6882	1.6901	1.6919	1.6938	1.6956	1.6974	1.6993	1.7011	1.7029	2	4	6	7	9	11	13	15	17

Natural Logarithms of 10^{+n}

n	1	2	3	4	5	6	7	8	9
ln 10^{+n}	2.3026	4.6052	6.9078	9.2103	11.5129	13.8155	16.1181	18.4207	20.7233

To find $\log_e 483.4$

$$483.4 = 4.834 \times 100 = 4.834 \times 10^2$$
$$\log_e 483.4 = \log_e 4.834 + \log_e 10^2$$
$$= 1.5756 + 4.6052 = 6.1808$$

Natural Logarithms

	0	1	2	3	4	5	6	7	8	9	1 2 3	4 5 6	7 8 9
5.5	1.7047	1.7066	1.7084	1.7102	1.7120	1.7138	1.7156	1.7174	1.7192	1.7210	2 4 5	7 9 11	13 14 16
5.6	1.7228	1.7246	1.7263	1.7281	1.7299	1.7317	1.7334	1.7352	1.7370	1.7387	2 4 5	7 9 11	12 14 16
5.7	1.7405	1.7422	1.7440	1.7457	1.7475	1.7492	1.7509	1.7527	1.7544	1.7561	2 3 5	7 9 10	12 14 16
5.8	1.7579	1.7596	1.7613	1.7630	1.7647	1.7664	1.7681	1.7699	1.7716	1.7733	2 3 5	7 9 10	12 14 15
5.9	1.7750	1.7766	1.7783	1.7800	1.7817	1.7834	1.7851	1.7867	1.7884	1.7901	2 3 5	7 8 10	12 13 15
6.0	1.7918	1.7934	1.7951	1.7967	1.7984	1.8001	1.8017	1.8034	1.8050	1.8066	2 3 5	7 8 10	12 13 15
6.1	1.8083	1.8099	1.8116	1.8132	1.8148	1.8165	1.8181	1.8197	1.8213	1.8229	2 3 5	7 8 10	11 13 15
6.2	1.8245	1.8262	1.8278	1.8294	1.8310	1.8326	1.8342	1.8358	1.8374	1.8390	2 3 5	6 8 10	11 13 14
6.3	1.8405	1.8421	1.8437	1.8453	1.8469	1.8485	1.8500	1.8516	1.8532	1.8547	2 3 5	6 8 9	11 13 14
6.4	1.8563	1.8579	1.8594	1.8610	1.8625	1.8641	1.8656	1.8672	1.8687	1.8703	2 3 5	6 8 9	11 12 14
6.5	1.8718	1.8733	1.8749	1.8764	1.8779	1.8795	1.8810	1.8825	1.8840	1.8856	2 3 5	6 8 9	11 12 14
6.6	1.8871	1.8886	1.8901	1.8916	1.8931	1.8946	1.8961	1.8976	1.8991	1.9006	2 3 5	6 8 9	11 12 14
6.7	1.9021	1.9036	1.9051	1.9066	1.9081	1.9095	1.9110	1.9125	1.9140	1.9155	1 3 4	6 7 9	10 12 13
6.8	1.9169	1.9184	1.9199	1.9213	1.9228	1.9242	1.9257	1.9272	1.9286	1.9301	1 3 4	6 7 9	10 12 13
6.9	1.9315	1.9330	1.9344	1.9359	1.9373	1.9387	1.9402	1.9416	1.9430	1.9445	1 3 4	6 7 9	10 12 13
7.0	1.9459	1.9473	1.9488	1.9502	1.9516	1.9530	1.9544	1.9559	1.9573	1.9587	1 3 4	6 7 9	10 11 13
7.1	1.9601	1.9615	1.9629	1.9643	1.9657	1.9671	1.9685	1.9699	1.9713	1.9727	1 3 4	6 7 8	10 11 13
7.2	1.9741	1.9755	1.9769	1.9782	1.9796	1.9810	1.9824	1.9838	1.9851	1.9865	1 3 4	6 7 8	10 11 12
7.3	1.9879	1.9892	1.9906	1.9920	1.9933	1.9947	1.9961	1.9974	1.9988	2.0001	1 3 4	5 7 8	10 11 12
7.4	2.0015	2.0028	2.0042	2.0055	2.0069	2.0082	2.0096	2.0109	2.0122	2.0136	1 3 4	5 7 8	9 11 12
7.5	2.0149	2.0162	2.0176	2.0189	2.0202	2.0215	2.0229	2.0242	2.0255	2.0268	1 3 4	5 7 8	9 11 12
7.6	2.0281	2.0295	2.0308	2.0321	2.0334	2.0347	2.0360	2.0373	2.0386	2.0399	1 3 4	5 7 8	9 10 12
7.7	2.0412	2.0425	2.0438	2.0451	2.0464	2.0477	2.0490	2.0503	2.0516	2.0528	1 3 4	5 6 8	9 10 12
7.8	2.0541	2.0554	2.0567	2.0580	2.0592	2.0605	2.0618	2.0631	2.0643	2.0656	1 3 4	5 6 8	9 10 11
7.9	2.0669	2.0681	2.0694	2.0707	2.0719	2.0732	2.0744	2.0757	2.0769	2.0782	1 3 4	5 6 8	9 10 11
8.0	2.0794	2.0807	2.0819	2.0832	2.0844	2.0857	2.0869	2.0882	2.0894	2.0906	1 2 4	5 6 7	9 10 11
8.1	2.0919	2.0931	2.0943	2.0956	2.0968	2.0980	2.0992	2.1005	2.1017	2.1029	1 2 4	5 6 7	9 10 11
8.2	2.1041	2.1054	2.1066	2.1078	2.1090	2.1102	2.1114	2.1126	2.1138	2.1150	1 2 4	5 6 7	8 10 11
8.3	2.1163	2.1175	2.1187	2.1199	2.1211	2.1223	2.1235	2.1247	2.1258	2.1270	1 2 4	5 6 7	8 10 11
8.4	2.1282	2.1294	2.1306	2.1318	2.1330	2.1342	2.1353	2.1365	2.1377	2.1389	1 2 4	5 6 7	8 9 11
8.5	2.1401	2.1412	2.1424	2.1436	2.1448	2.1459	2.1471	2.1483	2.1494	2.1506	1 2 4	5 6 7	8 9 11
8.6	2.1518	2.1529	2.1541	2.1552	2.1564	2.1576	2.1587	2.1599	2.1610	2.1622	1 2 3	5 6 7	8 9 10
8.7	2.1633	2.1645	2.1656	2.1668	2.1679	2.1691	2.1702	2.1713	2.1725	2.1736	1 2 3	5 6 7	8 9 10
8.8	2.1748	2.1759	2.1770	2.1782	2.1793	2.1804	2.1815	2.1827	2.1838	2.1849	1 2 3	5 6 7	8 9 10
8.9	2.1861	2.1872	2.1883	2.1894	2.1905	2.1917	2.1928	2.1939	2.1950	2.1961	1 2 3	4 6 7	8 9 10
9.0	2.1972	2.1983	2.1994	2.2006	2.2017	2.2028	2.2039	2.2050	2.2061	2.2072	1 2 3	4 6 7	8 9 10
9.1	2.2083	2.2094	2.2105	2.2116	2.2127	2.2138	2.2148	2.2159	2.2170	2.2181	1 2 3	4 5 7	8 9 10
9.2	2.2192	2.2203	2.2214	2.2225	2.2235	2.2246	2.2257	2.2268	2.2279	2.2289	1 2 3	4 5 6	8 9 10
9.3	2.2300	2.2311	2.2322	2.2332	2.2343	2.2354	2.2364	2.2375	2.2386	2.2396	1 2 3	4 5 6	7 9 10
9.4	2.2407	2.2418	2.2428	2.2439	2.2450	2.2460	2.2471	2.2481	2.2492	2.2502	1 2 3	4 5 6	7 8 10
9.5	2.2513	2.2523	2.2534	2.2544	2.2555	2.2565	2.2576	2.2586	2.2597	2.2607	1 2 3	4 5 6	7 8 9
9.6	2.2618	2.2628	2.2638	2.2649	2.2659	2.2670	2.2680	2.2690	2.2701	2.2711	1 2 3	4 5 6	7 8 9
9.7	2.2721	2.2732	2.2742	2.2752	2.2762	2.2773	2.2783	2.2793	2.2803	2.2814	1 2 3	4 5 6	7 8 9
9.8	2.2824	2.2834	2.2844	2.2854	2.2865	2.2875	2.2885	2.2895	2.2905	2.2915	1 2 3	4 5 6	7 8 9
9.9	2.2925	2.2935	2.2946	2.2956	2.2966	2.2976	2.2986	2.2996	2.3006	2.3016	1 2 3	4 5 6	7 8 9

Natural Logarithms of 10^{-n}

n	1	2	3	4	5	6	7	8	9
$\ln 10^{-n}$	$\bar{3}.6974$	$\bar{5}.3948$	$\bar{7}.0922$	$\bar{10}.7897$	$\bar{12}.4871$	$\bar{14}.1845$	$\bar{17}.8819$	$\bar{19}.5793$	$\bar{21}.2767$

To find $\log_e 0.053\,61$

$$0.053\,61 = \frac{5.361}{100} = \frac{5.361}{10^2} = 5.361 \times 10^{-2}$$

$$\log_e 0.05361 = \log_e 5.361 + \log_e 10^{-2}$$
$$= 1.6792 + \bar{5}.3948 = \bar{3}.0740$$
$$= -3 + 0.0740 = -2.9260$$

53

Table of e^x

x	.00	.01	.02	.03	.04	.05	.06	.07	.08	.09
0.0	1.0000	1.0101	1.0202	1.0305	1.0408	1.0513	1.0618	1.0725	1.0833	1.0942
0.1	1.1052	1.1163	1.1275	1.1388	1.1503	1.1618	1.1735	1.1853	1.1972	1.2092
0.2	1.2214	1.2337	1.2461	1.2586	1.2712	1.2840	1.2969	1.3100	1.3231	1.3364
0.3	1.3499	1.3634	1.3771	1.3910	1.4049	1.4191	1.4333	1.4477	1.4623	1.4770
0.4	1.4918	1.5068	1.5220	1.5373	1.5527	1.5683	1.5841	1.6000	1.6161	1.6323
0.5	1.6487	1.6653	1.6820	1.6989	1.7160	1.7333	1.7507	1.7683	1.7860	1.8040
0.6	1.8221	1.8404	1.8589	1.8776	1.8965	1.9155	1.9348	1.9542	1.9739	1.9937
0.7	2.0138	2.0340	2.0544	2.0751	2.0959	2.1170	2.1383	2.1598	2.1815	2.2034
0.8	2.2255	2.2479	2.2705	2.2933	2.3164	2.3396	2.3632	2.3869	2.4109	2.4351
0.9	2.4596	2.4843	2.5093	2.5345	2.5600	2.5857	2.6117	2.6379	2.6645	2.6912
1.0	2.7183	2.7456	2.7732	2.8011	2.8292	2.8576	2.8864	2.9154	2.9447	2.9743
1.1	3.0042	3.0344	3.0649	3.0957	3.1268	3.1582	3.1899	3.2220	3.2544	3.2871
1.2	3.3201	3.3535	3.3872	3.4212	3.4556	3.4903	3.5254	3.5608	3.5966	3.6328
1.3	3.6693	3.7062	3.7434	3.7810	3.8190	3.8574	3.8962	3.9354	3.9749	4.0149
1.4	4.0552	4.0960	4.1371	4.1787	4.2207	4.2631	4.3060	4.3492	4.3929	4.4371
1.5	4.4817	4.5267	4.5722	4.6182	4.6646	4.7115	4.7588	4.8066	4.8550	4.9037
1.6	4.9530	5.0028	5.0531	5.1039	5.1552	5.2070	5.2593	5.3122	5.3656	5.4195
1.7	5.4739	5.5290	5.5845	5.6407	5.6973	5.7546	5.8124	5.8709	5.9299	5.9895
1.8	6.0496	6.1104	6.1719	6.2339	6.2965	6.3598	6.4237	6.4883	6.5535	6.6194
1.9	6.6859	6.7531	6.8210	6.8895	6.9588	7.0287	7.0993	7.1707	7.2427	7.3155
2.0	7.3891	7.4633	7.5383	7.6141	7.6906	7.7679	7.8460	7.9248	8.0045	8.0849
2.1	8.1662	8.2482	8.3311	8.4149	8.4994	8.5849	8.6711	8.7583	8.8463	8.9352
2.2	9.0250	9.1157	9.2073	9.2999	9.3933	9.4877	9.5831	9.6794	9.7767	9.8749
2.3	9.9742	10.074	10.176	10.278	10.381	10.486	10.591	10.697	10.805	10.913
2.4	11.023	11.134	11.246	11.359	11.473	11.588	11.705	11.822	11.941	12.061
2.5	12.183	12.305	12.429	12.554	12.680	12.807	12.936	13.066	13.197	13.330
2.6	13.464	13.599	13.736	13.874	14.013	14.154	14.296	14.440	14.585	14.732
2.7	14.880	15.029	15.180	15.333	15.487	15.643	15.800	15.959	16.119	16.281
2.8	16.445	16.610	16.777	16.945	17.116	17.288	17.462	17.637	17.814	17.993
2.9	18.174	18.357	18.541	18.728	18.916	19.106	19.298	19.492	19.688	19.886
3.0	20.086	20.287	20.491	20.697	20.905	21.115	21.327	21.542	21.758	21.977
3.1	22.198	22.421	22.646	22.874	23.104	23.336	23.571	23.808	24.047	24.288
3.2	24.533	24.779	25.028	25.280	25.534	25.790	26.050	26.311	26.576	26.843
3.3	27.113	27.385	27.660	27.938	28.219	28.503	28.789	29.079	29.371	29.666
3.4	29.964	30.265	30.569	30.877	31.187	31.500	31.817	32.137	32.460	32.786
3.5	33.115	33.448	33.784	34.124	34.467	34.813	35.163	35.517	35.874	36.234
3.6	36.598	36.966	37.338	37.713	38.092	38.475	38.861	39.252	39.646	40.045
3.7	40.447	40.854	41.264	41.679	42.098	42.521	42.948	43.380	43.816	44.256
3.8	44.701	45.150	45.604	46.063	46.525	46.993	47.465	47.942	48.424	48.911
3.9	49.402	49.899	50.400	50.907	51.419	51.935	52.457	52.985	53.517	54.055
4.0	54.598									

x	e^x	x	e^x	x	e^x	x	e^x	x	e^x	x	e^x
4.1	60.340	4.6	99.484	5.1	164.02	5.6	270.43	6.1	445.86	6.6	735.10
4.2	66.686	4.7	109.95	5.2	181.27	5.7	298.87	6.2	492.75	6.7	812.41
4.3	73.700	4.8	121.51	5.3	200.34	5.8	330.30	6.3	544.57	6.8	897.85
4.4	81.451	4.9	134.29	5.4	221.41	5.9	365.04	6.4	601.85	6.9	992.27
4.5	90.017	5.0	148.41	5.5	244.69	6.0	403.43	6.5	665.14	7.0	1096.63

Table of e^{-x}

x	.00	.01	.02	.03	.04	.05	.06	.07	.08	.09
0.0	1.0000	0.9900	0.9802	0.9704	0.9608	0.9512	0.9418	0.9324	0.9231	0.9139
0.1	0.9048	0.8958	0.8869	0.8781	0.8694	0.8607	0.8521	0.8437	0.8353	0.8270
0.2	0.8187	0.8106	0.8025	0.7945	0.7866	0.7788	0.7711	0.7634	0.7558	0.7483
0.3	0.7408	0.7334	0.7261	0.7189	0.7118	0.7047	0.6977	0.6907	0.6839	0.6771
0.4	0.6703	0.6637	0.6570	0.6505	0.6440	0.6376	0.6313	0.6250	0.6188	0.6126
0.5	0.6065	0.6005	0.5945	0.5886	0.5827	0.5769	0.5712	0.5655	0.5599	0.5543
0.6	0.5488	0.5434	0.5379	0.5326	0.5273	0.5220	0.5169	0.5117	0.5066	0.5016
0.7	0.4966	0.4916	0.4868	0.4819	0.4771	0.4724	0.4677	0.4630	0.4584	0.4538
0.8	0.4493	0.4449	0.4404	0.4360	0.4317	0.4274	0.4232	0.4190	0.4148	0.4107
0.9	0.4066	0.4025	0.3985	0.3946	0.3906	0.3867	0.3829	0.3791	0.3753	0.3716
1.0	0.3679	0.3642	0.3606	0.3570	0.3535	0.3499	0.3465	0.3430	0.3396	0.3362
1.1	0.3329	0.3296	0.3263	0.3230	0.3198	0.3166	0.3135	0.3104	0.3073	0.3042
1.2	0.3012	0.2982	0.2952	0.2923	0.2894	0.2865	0.2837	0.2808	0.2780	0.2753
1.3	0.2725	0.2698	0.2671	0.2645	0.2618	0.2592	0.2567	0.2541	0.2516	0.2491
1.4	0.2466	0.2441	0.2417	0.2393	0.2369	0.2346	0.2322	0.2299	0.2276	0.2254
1.5	0.2231	0.2209	0.2187	0.2165	0.2144	0.2122	0.2101	0.2080	0.2060	0.2039
1.6	0.2019	0.1999	0.1979	0.1959	0.1940	0.1920	0.1901	0.1882	0.1864	0.1845
1.7	0.1827	0.1809	0.1791	0.1773	0.1755	0.1738	0.1720	0.1703	0.1686	0.1670
1.8	0.1653	0.1637	0.1620	0.1604	0.1588	0.1572	0.1557	0.1541	0.1526	0.1511
1.9	0.1496	0.1481	0.1466	0.1451	0.1437	0.1423	0.1409	0.1395	0.1381	0.1367
2.0	0.1353	0.1340	0.1327	0.1313	0.1300	0.1287	0.1275	0.1262	0.1249	0.1237
2.1	0.1225	0.1212	0.1200	0.1188	0.1177	0.1165	0.1153	0.1142	0.1130	0.1119
2.2	0.1108	0.1097	0.1086	0.1075	0.1065	0.1054	0.1044	0.1033	0.1023	0.1013
2.3	0.1003	0.0993	0.0983	0.0973	0.0963	0.0954	0.0944	0.0935	0.0925	0.0916
2.4	0.0907	0.0898	0.0889	0.0880	0.0872	0.0863	0.0854	0.0846	0.0837	0.0829
2.5	0.0821	0.0813	0.0805	0.0797	0.0789	0.0781	0.0773	0.0765	0.0758	0.0750
2.6	0.0743	0.0735	0.0728	0.0721	0.0714	0.0707	0.0699	0.0693	0.0686	0.0679
2.7	0.0672	0.0665	0.0659	0.0652	0.0646	0.0639	0.0633	0.0627	0.0620	0.0614
2.8	0.0608	0.0602	0.0596	0.0590	0.0584	0.0578	0.0573	0.0567	0.0561	0.0556
2.9	0.0550	0.0545	0.0539	0.0534	0.0529	0.0523	0.0518	0.0513	0.0508	0.0503
3.0	0.0498	0.0493	0.0488	0.0483	0.0478	0.0474	0.0469	0.0464	0.0460	0.0455
3.1	0.0450	0.0446	0.0442	0.0437	0.0433	0.0429	0.0424	0.0420	0.0416	0.0412
3.2	0.0408	0.0404	0.0400	0.0396	0.0392	0.0388	0.0384	0.0380	0.0376	0.0373
3.3	0.0369	0.0365	0.0362	0.0358	0.0354	0.0351	0.0347	0.0344	0.0340	0.0337
3.4	0.0334	0.0330	0.0327	0.0324	0.0321	0.0317	0.0314	0.0311	0.0308	0.0305
3.5	0.0302	0.0299	0.0296	0.0293	0.0290	0.0287	0.0284	0.0282	0.0279	0.0276
3.6	0.0273	0.0271	0.0268	0.0265	0.0263	0.0260	0.0257	0.0255	0.0252	0.0250
3.7	0.0247	0.0245	0.0242	0.0240	0.0238	0.0235	0.0233	0.0231	0.0228	0.0226
3.8	0.0224	0.0221	0.0219	0.0217	0.0215	0.0213	0.0211	0.0209	0.0207	0.0204
3.9	0.0202	0.0200	0.0198	0.0196	0.0194	0.0193	0.0191	0.0189	0.0187	0.0185
4.0	0.0183									

e^{-x} for values of x greater than 4.0 may be found by using the table of natural logarithms and the reciprocal table.

To find $e^{-4.5361}$

If $y = e^{4.5361}$ then from the e^x table y is about 90.
Since 4.5361 is outside the range of values given in the natural logarithm tables, we use the table of natural logarithms of 10^{+n}
We find that $\log_e 10 = 2.3026$
and hence $\log_e y = 2.3026 + 2.335$
$\qquad \log_e 9.333 = 2.2335$ (using the natural logarithm table)
Hence $e^{4.5361} = 9.333 \times 10 = 93.33$
$\qquad e^{-4.5361} = 0.01072$ (using the table of reciprocals)

Normal Distribution

u	0.00	0.01	0.02	0.03	0.04	0.05	0.06	0.07	0.08	0.09
0.0	0.50000	0.49601	0.49202	0.48803	0.48405	0.48006	0.47608	0.47210	0.46812	0.46414
0.1	0.46017	0.45620	0.45224	0.44828	0.44433	0.44038	0.43644	0.43250	0.42858	0.42465
0.2	0.42074	0.41683	0.41294	0.40905	0.40517	0.40129	0.39743	0.39358	0.38974	0.38591
0.3	0.38209	0.37828	0.37448	0.37070	0.36693	0.36317	0.35942	0.35569	0.35197	0.34827
0.4	0.34458	0.34090	0.33724	0.33360	0.32997	0.32636	0.32276	0.31918	0.31561	0.31207
0.5	0.30854	0.30503	0.30153	0.29806	0.29460	0.29116	0.28774	0.28434	0.28096	0.27760
0.6	0.27425	0.27093	0.26763	0.26435	0.26109	0.25785	0.25463	0.25143	0.24825	0.24510
0.7	0.24196	0.23885	0.23576	0.23269	0.22965	0.22663	0.22363	0.22065	0.21770	0.21476
0.8	0.21186	0.20897	0.20611	0.20327	0.20045	0.19766	0.19489	0.19215	0.18943	0.18673
0.9	0.18406	0.18141	0.17879	0.17619	0.17361	0.17106	0.16853	0.16602	0.16354	0.16109
1.0	0.15866	0.15625	0.15386	0.15150	0.14917	0.14686	0.14457	0.14231	0.14007	0.13786
1.1	0.13567	0.13350	0.13136	0.12924	0.12714	0.12507	0.12302	0.12100	0.11900	0.11702
1.2	0.11507	0.11314	0.11123	0.10935	0.10749	0.10565	0.10383	0.10204	0.10027	0.09853
1.3	0.09680	0.09510	0.09342	0.09176	0.09012	0.08851	0.08692	0.08534	0.08379	0.08226
1.4	0.08076	0.07927	0.07780	0.07636	0.07493	0.07353	0.07215	0.07078	0.06944	0.06811
1.5	0.06681	0.06552	0.06426	0.06301	0.06178	0.06057	0.05938	0.05821	0.05705	0.05592
1.6	0.05480	0.05370	0.05262	0.05155	0.05050	0.04947	0.04846	0.04746	0.04648	0.04551
1.7	0.04457	0.04363	0.04272	0.04182	0.04093	0.04006	0.03920	0.03836	0.03754	0.03673
1.8	0.03593	0.03515	0.03438	0.03362	0.03288	0.03216	0.03144	0.03074	0.03005	0.02938
1.9	0.02872	0.02807	0.02743	0.02680	0.02619	0.02559	0.02500	0.02442	0.02385	0.02330
2.0	0.02275	0.02222	0.02169	0.02118	0.02068	0.02018	0.01970	0.01923	0.01876	0.01831
2.1	0.01786	0.01743	0.01700	0.01659	0.01618	0.01578	0.01539	0.01500	0.01463	0.01426
2.2	0.01390	0.01355	0.01321	0.01287	0.01255	0.01222	0.01191	0.01160	0.01130	0.01101
2.3	0.01072	0.01044	0.01017	0.00990	0.00964	0.00939	0.00914	0.00889	0.00866	0.00842
2.4	0.00820	0.00798	0.00776	0.00755	0.00734	0.00714	0.00695	0.00676	0.00657	0.00639
2.5	0.00621	0.00604	0.00587	0.00570	0.00554	0.00539	0.00523	0.00508	0.00494	0.00480
2.6	0.00466	0.00453	0.00440	0.00427	0.00415	0.00402	0.00391	0.00379	0.00368	0.00357
2.7	0.00347	0.00336	0.00326	0.00317	0.00307	0.00298	0.00289	0.00280	0.00272	0.00264
2.8	0.00256	0.00248	0.00240	0.00233	0.00226	0.00219	0.00212	0.00205	0.00199	0.00193
2.9	0.00187	0.00181	0.00175	0.00169	0.00164	0.00159	0.00154	0.00149	0.00144	0.00139
3.0	0.00135	0.00131	0.00126	0.00122	0.00118	0.00114	0.00111	0.00107	0.00104	0.00100
3.1	0.00097	0.00094	0.00090	0.00087	0.00084	0.00082	0.00079	0.00076	0.00074	0.00071
3.2	0.00069	0.00066	0.00064	0.00062	0.00060	0.00058	0.00056	0.00054	0.00052	0.00050
3.3	0.00048	0.00047	0.00045	0.00043	0.00042	0.00040	0.00039	0.00038	0.00036	0.00035
3.4	0.00034	0.00032	0.00031	0.00030	0.00029	0.00028	0.00027	0.00026	0.00025	0.00024
3.5	0.00023	0.00022	0.00022	0.00021	0.00020	0.00019	0.00019	0.00018	0.00017	0.00017
3.6	0.00016	0.00015	0.00015	0.00014	0.00014	0.00013	0.00013	0.00012	0.00012	0.00011
3.7	0.00011	0.00010	0.00010	0.00010	0.00009	0.00009	0.00008	0.00008	0.00008	0.00008
3.8	0.00007	0.00007	0.00007	0.00006	0.00006	0.00006	0.00006	0.00005	0.00005	0.00005
3.9	0.00005	0.00005	0.00004	0.00004	0.00004	0.00004	0.00004	0.00004	0.00003	0.00003

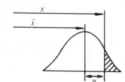

$$u = \frac{x - \bar{x}}{\sigma}$$

x = value of the variate
$\bar{x}$ = arithmetic mean
σ = standard deviation

Binomial Coefficients

r:	0	1	2	3	4	5	6	7	8	9	10
$n=$											
1	1	1									
2	1	2	1								
3	1	3	3	1							
4	1	4	6	4	1						
5	1	5	10	10	5	1					
6	1	6	15	20	15	6	1				
7	1	7	21	35	35	21	7	1			
8	1	8	28	56	70	56	28	8	1		
9	1	9	36	84	126	126	84	36	9	1	
10	1	10	45	120	210	252	210	120	45	10	1
11	1	11	55	165	330	462	462	330	165	55	11
12	1	12	66	220	495	792	924	792	495	220	66
13	1	13	78	286	715	1287	1716	1716	1287	715	286
14	1	14	91	364	1001	2002	3003	3432	3003	2002	1001
15	1	15	105	455	1365	3003	5005	6435	6435	5005	3003
16	1	16	120	560	1820	4368	8008	11440	12870	11440	8008
17	1	17	136	680	2380	6188	12376	19448	24310	24310	19448
18	1	18	153	816	3060	8568	18564	31824	43758	48620	43758
19	1	19	171	969	3876	11628	27132	50388	75582	92378	92378
20	1	20	190	1140	4845	15504	38760	77520	125970	167960	184756

$(a+b)^5 = a^5 + 5a^4b + 10a^3b^2 + 10a^2b^3 + 5ab^4 + b^5.$

Note that the coefficients are symmetrical. Thus

$$(a+b)^{15} = a^{15} + 15a^{14}b + 105a^{13}b^2 + 455a^{12}b^3 + 1365a^{11}b^4 + 3003a^{10}b^5$$
$$+ 5005a^9b^6 + 6435a^8b^7 + 6435a^7b^8 + 5005a^6b^9 + 3003a^5b^{10}$$
$$+ 1365a^4b^{11} + 455a^3b^{12} + 105a^2b^{13} + 15ab^{14} + b^{15}.$$

Binomial distribution

If the probability of finding a defective item in a single trial is p then the probability of finding 0, 1, 2, 3 ... defective items in a sample of n items is the successive terms of the expansion of $(q+p)^n$, where $q = 1-p$. Note that p is the fraction of defective items produced by the process.

Number of defective items in the sample	0	1	2
Probability	q^n	$nq^{n-1}p$	$\dfrac{n(n-1)}{2!}q^{n-2}p^2$

Approximations to the binomial distribution

(i) If np is less than 5 then the binomial distribution is well approximated by the Poisson distribution with $\lambda = np$.

(ii) If np is greater than 5 then the binomial distribution may be approximated by a normal distribution with $\bar{x} = np$ and $\sigma = \sqrt{npq}$

Poisson Distribution

x	0.1	0.2	0.3	0.4	0.5	0.6	0.7	0.8	0.9	1.0
0	0.9048	0.8187	0.7408	0.6703	0.6065	0.5488	0.4966	0.4493	0.4066	0.3679
1	0.0905	0.1637	0.2222	0.2681	0.3033	0.3293	0.3476	0.3595	0.3659	0.3679
2	0.0045	0.0164	0.0333	0.0536	0.0758	0.0988	0.1217	0.1438	0.1647	0.1839
3	0.0002	0.0011	0.0033	0.0072	0.0126	0.0198	0.0284	0.0383	0.0494	0.0613
4	0.0000	0.0001	0.0002	0.0007	0.0016	0.0030	0.0050	0.0077	0.0111	0.0153
5	0.0000	0.0000	0.0000	0.0001	0.0002	0.0004	0.0007	0.0012	0.0020	0.0031
6	0.0000	0.0000	0.0000	0.0000	0.0000	0.0000	0.0001	0.0002	0.0003	0.0005
7	0.0000	0.0000	0.0000	0.0000	0.0000	0.0000	0.0000	0.0000	0.0000	0.0001

x	1.1	1.2	1.3	1.4	1.5	1.6	1.7	1.8	1.9	2.0
0	0.3329	0.3012	0.2725	0.2466	0.2231	0.2019	0.1827	0.1653	0.1496	0.1353
1	0.3662	0.3614	0.3543	0.3452	0.3347	0.3230	0.3106	0.2975	0.2842	0.2707
2	0.2014	0.2169	0.2303	0.2417	0.2510	0.2584	0.2640	0.2678	0.2700	0.2707
3	0.0738	0.0867	0.0998	0.1128	0.1255	0.1378	0.1496	0.1607	0.1710	0.1804
4	0.0203	0.0260	0.0324	0.0395	0.0471	0.0551	0.0636	0.0723	0.0812	0.0902
5	0.0045	0.0062	0.0084	0.0111	0.0141	0.0176	0.0216	0.0260	0.0309	0.0361
6	0.0008	0.0012	0.0018	0.0026	0.0035	0.0047	0.0061	0.0078	0.0098	0.0120
7	0.0001	0.0002	0.0003	0.0005	0.0008	0.0011	0.0015	0.0020	0.0027	0.0034
8	0.0000	0.0000	0.0001	0.0001	0.0001	0.0002	0.0003	0.0005	0.0006	0.0009
9	0.0000	0.0000	0.0000	0.0000	0.0000	0.0000	0.0001	0.0001	0.0001	0.0002

x	2.1	2.2	2.3	2.4	2.5	2.6	2.7	2.8	2.9	3.0
0	0.1225	0.1108	0.1003	0.0907	0.0821	0.0743	0.0672	0.0608	0.0550	0.0498
1	0.2572	0.2438	0.2306	0.2177	0.2052	0.1931	0.1815	0.1703	0.1596	0.1494
2	0.2700	0.2681	0.2652	0.2613	0.2565	0.2510	0.2450	0.2384	0.2314	0.2240
3	0.1890	0.1966	0.2033	0.2090	0.2138	0.2176	0.2205	0.2225	0.2237	0.2240
4	0.0992	0.1082	0.1169	0.1254	0.1336	0.1414	0.1488	0.1557	0.1622	0.1680
5	0.0417	0.0476	0.0538	0.0602	0.0668	0.0735	0.0804	0.0872	0.0940	0.1008
6	0.0146	0.0174	0.0206	0.0241	0.0278	0.0319	0.0362	0.0407	0.0455	0.0504
7	0.0044	0.0055	0.0068	0.0083	0.0099	0.0118	0.0139	0.0163	0.0188	0.0216
8	0.0011	0.0015	0.0019	0.0025	0.0031	0.0038	0.0047	0.0057	0.0068	0.0081
9	0.0003	0.0004	0.0005	0.0007	0.0009	0.0011	0.0014	0.0018	0.0022	0.0027
10	0.0001	0.0001	0.0001	0.0002	0.0002	0.0003	0.0004	0.0005	0.0006	0.0008
11	0.0000	0.0000	0.0000	0.0000	0.0000	0.0001	0.0001	0.0001	0.0002	0.0002
12	0.0000	0.0000	0.0000	0.0000	0.0000	0.0000	0.0000	0.0000	0.0000	0.0001

Entries in the table give the probabilities that an event will occur x times when the average number of occurrences is λ.

Poisson Distribution

χ	3.1	3.2	3.3	3.4	3.5	3.6	3.7	3.8	3.9	4.0
0	0.0450	0.0408	0.0369	0.0334	0.0302	0.0273	0.0247	0.0224	0.0202	0.0183
1	0.1397	0.1304	0.1217	0.1135	0.1057	0.0984	0.0915	0.0850	0.0789	0.0733
2	0.2165	0.2087	0.2008	0.1929	0.1850	0.1771	0.1692	0.1615	0.1539	0.1465
3	0.2237	0.2226	0.2209	0.2186	0.2158	0.2125	0.2087	0.2046	0.2001	0.1954
4	0.1734	0.1781	0.1823	0.1858	0.1888	0.1912	0.1931	0.1944	0.1951	0.1954
5	0.1075	0.1140	0.1203	0.1264	0.1322	0.1377	0.1429	0.1477	0.1522	0.1563
6	0.0555	0.0608	0.0662	0.0716	0.0771	0.0826	0.0881	0.0936	0.0989	0.1042
7	0.0246	0.0278	0.0312	0.0348	0.0385	0.0425	0.0466	0.0508	0.0551	0.0595
8	0.0095	0.0111	0.0129	0.0148	0.0169	0.0191	0.0215	0.0241	0.0269	0.0298
9	0.0033	0.0040	0.0047	0.0056	0.0066	0.0076	0.0089	0.0102	0.0116	0.0132
10	0.0010	0.0013	0.0016	0.0019	0.0023	0.0028	0.0033	0.0039	0.0045	0.0053
11	0.0003	0.0004	0.0005	0.0006	0.0007	0.0009	0.0011	0.0013	0.0016	0.0019
12	0.0001	0.0001	0.0001	0.0002	0.0002	0.0003	0.0003	0.0004	0.0005	0.0006
13	0.0000	0.0000	0.0000	0.0000	0.0001	0.0001	0.0001	0.0001	0.0002	0.0002
14	0.0000	0.0000	0.0000	0.0000	0.0000	0.0000	0.0000	0.0000	0.0000	0.0001

χ	4.1	4.2	4.3	4.4	4.5	4.6	4.7	4.8	4.9	5.0
0	0.0166	0.0150	0.0136	0.0123	0.0111	0.0101	0.0091	0.0082	0.0074	0.0067
1	0.0679	0.0630	0.0583	0.0540	0.0500	0.0462	0.0427	0.0395	0.0365	0.0337
2	0.1393	0.1323	0.1254	0.1188	0.1125	0.1063	0.1005	0.0948	0.0894	0.0842
3	0.1904	0.1852	0.1798	0.1743	0.1687	0.1631	0.1574	0.1517	0.1460	0.1404
4	0.1951	0.1944	0.1933	0.1917	0.1898	0.1875	0.1849	0.1820	0.1789	0.1755
5	0.1600	0.1633	0.1662	0.1687	0.1708	0.1725	0.1738	0.1747	0.1753	0.1755
6	0.1093	0.1143	0.1191	0.1237	0.1281	0.1323	0.1362	0.1398	0.1432	0.1462
7	0.0640	0.0686	0.0732	0.0778	0.0824	0.0869	0.0914	0.0959	0.1002	0.1044
8	0.0328	0.0360	0.0393	0.0428	0.0463	0.0500	0.0537	0.0575	0.0614	0.0653
9	0.0150	0.0168	0.0188	0.0209	0.0232	0.0255	0.0280	0.0307	0.0334	0.0363
10	0.0061	0.0071	0.0081	0.0092	0.0104	0.0118	0.0132	0.0147	0.0164	0.0181
11	0.0023	0.0027	0.0032	0.0037	0.0043	0.0049	0.0056	0.0064	0.0073	0.0082
12	0.0008	0.0009	0.0011	0.0014	0.0016	0.0019	0.0022	0.0026	0.0030	0.0034
13	0.0002	0.0003	0.0004	0.0005	0.0006	0.0007	0.0008	0.0009	0.0011	0.0013
14	0.0001	0.0001	0.0001	0.0001	0.0002	0.0002	0.0003	0.0003	0.0004	0.0005
15	0.0000	0.0000	0.0000	0.0000	0.0001	0.0001	0.0001	0.0001	0.0001	0.0002

Entries in the table give the probabilities that an event will occur χ times when the average number of occurrences is λ.

Poisson distribution

If λ is the expected number of defective items in a sample then the probabilities of finding $0, 1, 2, 3 \ldots$ defective items in the sample is given by the successive terms of the expansion of $e^{-\lambda}e^{\lambda}$.

Number of defective items in the sample	0	1	2
Probability	$e^{-\lambda}$	$\lambda e^{-\lambda}$	$\dfrac{\lambda^2}{2!} e^{-\lambda}$

Statistical Formulae

Control chart for sample means

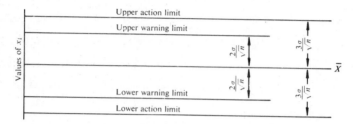

$$\bar{X} = \frac{\bar{x}_1 + \bar{x}_2 + \bar{x}_3 + \dots + \bar{x}_n}{n}$$ where $\bar{X}$ = overall mean and $\bar{x}_i$ = mean of ith sample

$$\bar{R} = \frac{R_1 + R_2 + R_3 + \dots + R_n}{n}$$ where $\bar{R}$ = average range and R_i = range of ith sample

$\sigma = d\bar{R}$ where d is a constant which depends upon the sample size.

Values of d

Sample Size	d	Sample Size	d	Sample Size	d	Sample Size	d	Sample Size	d
2	0.8862	6	0.3946	10	0.3249	14	0.2935	18	0.2747
3	0.5908	7	0.3698	11	0.3152	15	0.2880	19	0.2711
4	0.4857	8	0.3512	12	0.3069	16	0.2831	20	0.2677
5	0.4299	9	0.3367	13	0.2998	17	0.2787		

Warning limits $= \bar{X} \pm \frac{2\sigma}{\sqrt{n}}$ Action limits $= \bar{X} \pm \frac{3\sigma}{\sqrt{n}}$

The warning and action limits can also be calculated as follows:

Warning limits $= \bar{X} \pm A_{0.025} \times \bar{R}$ Action limits $= \bar{X} \pm A_{0.001} \times \bar{R}$

where $\bar{R}$ is the average range of the samples and A is a constant depending upon the sample size.

Values of A

Sample size	$A_{0.025}$	$A_{0.001}$	Sample size	$A_{0.025}$	$A_{0.001}$
2	1.23	1.94	7	0.27	0.43
3	0.67	1.05	8	0.24	0.38
4	0.48	0.75	9	0.22	0.35
5	0.38	0.59	10	0.20	0.32
6	0.32	0.50			

Note that $A_{0.025} \times \bar{R} = \frac{2\sigma}{\sqrt{n}}$ and that $A_{0.001} \times \bar{R} = \frac{3\sigma}{\sqrt{n}}$ (approximately)

The range chart

The control limits for the range chart are obtained by using the following:

upper action limit $= D'_{0.999} \times \bar{R}$ lower warning limit $= D'_{0.025} \times \bar{R}$

upper warning limit $= D'_{0.975} \times \bar{R}$ lower action limit $= D'_{0.001} \times \bar{R}$

D' is a constant depending upon the sample size

Values of D'

Sample size	$D'_{0.001}$	$D'_{0.025}$	$D'_{0.975}$	$D'_{0.999}$	Sample size	$D'_{0.001}$	$D'_{0.025}$	$D'_{0.975}$	$D'_{0.999}$
2	0.01	0.04	2.81	4.12	7	0.26	0.46	1.66	2.11
3	0.04	0.18	2.17	2.98	8	0.29	0.50	1.62	2.04
4	0.10	0.29	1.93	2.58	9	0.32	0.52	1.58	1.99
5	0.16	0.37	1.81	2.36	10	0.35	0.54	1.56	1.93
6	0.21	0.42	1.72	2.22					

Tests for an assignable cause: (i) One point falling outside the action limit.

                                       (ii) Two consecutive points falling outside the warning limit.

                                       (iii) Three consecutive points falling outside the limits $\bar{X} \pm \sigma$.

                                       (iv) Eleven consecutive points falling on the same side of the overall mean value.

Number defective control chart

$\bar{X} = np$ and $\sigma = \sqrt{npq}$

$$p = \frac{\text{total number of defectives in all the samples checked}}{\text{total number of items checked}}$$

$q = 1 - p$

Action limits $= \bar{X} \pm 3\sigma$

Warning limits $= \bar{X} \pm 2\sigma$

Percentage defective control chart

$\bar{X} = p$ and $\sigma = \sqrt{\dfrac{pq}{n}}$

Action limits $= p \pm 3\sigma$

Warning limits $= p \pm 2\sigma$

where $p = $ percentage defective and $q = 100 - p$

Chart for average number of defects per item

$$z = \frac{\text{total number of defects observed}}{\text{total number of items checked}}$$

$\bar{X} = z$ and $\sigma = \sqrt{\dfrac{z}{n}}$

Action limits $= z \pm 3\sigma$

Warning limits $= z \pm 2\sigma$

Compressed limit control charts

The sampling gauges are set at $0.2 \times$ gross tolerance inside the specification limits. The table below gives the sample sizes recommended together with the corresponding action limit.

Sample size	15	20	25
Action limit	6	7	8

Statistical Formulae

Arithmetic mean $\quad \bar{x} = \dfrac{\Sigma fx}{N}$

x = variable
f = frequency
N = total frequency

Standard deviation $\quad \sigma = \sqrt{\dfrac{\Sigma f(x-\bar{x})^2}{N}} = \sqrt{\dfrac{\Sigma fx^2}{N} - \bar{x}^2}$

Coded method for calculating the mean and standard deviation:

$\bar{x}$ = assumed mean $+\bar{x}_c \times$ unit size $\qquad \bar{x}_c = \dfrac{\Sigma fx_c}{N}$

$\sigma = \sigma_c \times$ unit size $\qquad\qquad\qquad \sigma_c = \sqrt{\dfrac{\Sigma fx_c^2}{N} - \bar{x}_c^2}$

Example

Class	x	x_c	f	fx_c	$fx^2{}_c$
5.94 – 5.96	5.95	−5	8	−40	200
5.97 – 5.99	5.98	−2	37	−74	148
6.00 – 6.02	6.01	1	90	90	90
6.03 – 6.05	6.04	4	52	208	832
6.06 – 6.08	6.07	7	13	91	637
			200	275	1907

$\bar{x}_c = \dfrac{275}{200} = 1.375$

$\bar{x} = 6.00 + 1.375 \times 0.01$
$\quad = 6.00 + 0.01375 = 6.013\,75$

$\sigma_c = \sqrt{\dfrac{1907}{200} - (1.375)^2}$

$\quad = \sqrt{9.535 - 1.891}$
$\quad = \sqrt{7.644} = 2.765$
$\sigma = 2.765 \times 0.01 = 0.027\,65$

assumed mean = 6.00
unit size = 0.01

Simple probability $\quad p = \dfrac{\text{number of ways in which an event can happen}}{\text{total number of ways which are possible}}$

To find the probability of throwing a six on a die:

number of ways in which a six can be thrown $\Rightarrow 1$

total number of ways that are possible $\quad = 6$

$$p = \frac{1}{6}$$

The value of a probability always lies between 0 and 1.

Linear correlation

The least square line approximating to the set of points $(x_1, y_1), (x_2, y_2) \ldots (x_n, y_n)$ has the equation $y = a + bx$. The constants a and b are found by solving the simultaneous equations:

$$\Sigma y = aN + b\ \Sigma x \ldots (1)$$
$$\Sigma xy = a\ \Sigma x + b\ \Sigma x^2 \ldots (2)$$

where N is the number of points.
The coefficient of correlation is

$$r = \frac{\Sigma XY}{\sqrt{\Sigma X^2\ \Sigma Y^2}} \text{ where } X = x - \bar{x} \text{ and } Y = y - \bar{y}$$

For perfect positive correlation $r = +1$
For perfect negative correlation $r = -1$

Physics Data

Heat

Expansion of solids:

$$l = l_o(l + at)$$

l = final length
l_o = original length
a = coefficient of linear expansion
t = temperature change

Expansion of liquids and gases:

$$V = V_o(1 + \beta t)$$

V_o = original volume
V = final volume
β = coefficient of cubical expansion

Boyle's Law:

$$p_1 V_1 = p_2 V_2$$

p = pressure (N/m^2, Pa)

Charle's Law:

$$\frac{V_1}{T_1} = \frac{V_2}{T_2}$$

T = temperature (K)

Combination of Boyle's and Charle's Law:

$$\frac{p_1 V_1}{T_1} = \frac{p_2 V_2}{T_2}$$

m = mass
R = gas constant
C = a constant
c_p = specific heat capacity at constant pressure
c_v = specific heat capacity at constant volume

Gas characteristic equation:

$$pV = mRT$$

Adiabatic expansion:

$$pV^\gamma = C$$

$$\gamma = \frac{c_P}{c_V}$$

h_g = enthalpy of dry stream (J/kg)
h_f = liquid enthalpy of stream (J/kg)
h_{fg} = enthalpy of evaporation of stream (J/kg)
x = dryness fraction

Polytropic expansion:

$$pV^n = C$$

Enthalpy of steam:

$$h_g = h_f + h_{fg}$$
$$h = h_f + xh_{fg}$$

Sound

$$v = f\lambda$$

v = velocity (m/s)
f = frequency (Hz)
λ = wavelength (m)

Velocity of sound in dry air at $0°C = 331$ m/s

Light

Refraction:

$$n = \frac{\sin i}{\sin r}$$

n = refractive index
i = angle of incidence
r = angle of refraction

$$n = \frac{\text{real depth}}{\text{apparent depth}}$$

Lenses: The focal length of a lens is the distance between the optical centre and the principal focus.

$$m = \frac{v}{u}$$

$$\frac{1}{v} - \frac{1}{u} = \frac{1}{f}$$

m = linear magnification
v = distance of image from lens
u = distance of object from lens

Properties of common metals

Metal	Melting point °C	Density kg/m³	E GN/m² or GPa	G GN/m² or GPa	Relative specific heat capacity	Coefft. of linear expansion $\times 10^{-6}$/°C	Resistivity at 0°C $\mu\Omega$m	Resistance temp. coeff. at 0°C mΩ/°C	Electro-chemical equivalent mg/C
Aluminium	659	2700	70	27	0.21	23	245	450	0.093
Copper	1083	8900	96	38	0.09	17	156	430	0.329
Gold	1063	19300	79	27	0.03	14	204	400	0.681
Iron	1475	7850	200	82	0.11	12	890	650	0.193
Lead	327	11370	16	—	0.03	29	1900	420	1.074
Mercury	—	13580	—	—	0.03	60	9410	100	1.039
Nickel	1452	8800	198	—	0.11	13	614	680	0.304
Platinum	1775	21040	164	51	0.03	9	981	390	0.506
Silver	961	10530	78	29	0.06	19	151	410	1.118
Tungsten	3400	19300	410	—	0.03	4.5	490	480	0.318
Zinc	419	6860	86	38	0.09	30	550	420	0.339

Properties of some copper alloys

Names and uses	Composition %			Condition	Mechanical properties			
	Cu	Zn	Others		0.1% P.S. N/mm² or MPa	T N/mm² or MPa	Elong. %	Vickers Hardness
Muntz metal for die stamping and extrusions	60	40	—	Extruded	110	350	40	75
Free cutting brass for high speed machining	58	39	Lead 3%	Extruded	140	440	30	100
Cartridge brass for severe cold working	70	30	—	Annealed	75	270	70	65
				Work hardened	500	600	5	180
Standard brass for press work	65	35	—	Annealed	90	320	65	65
				Work hardened	500	690	4	185
Admiralty gun-metal for general purpose castings	88	2	Tin 10%	Sand casting	120	290	16	85
Phosphor bronze for castings and bushes for bearings	Rem.		Tin 10% Phosphorus 0.03-0.25%	Sand casting	120	280	15	90

Cu = copper   Zn = zinc

T = tensile strength

Properties of cast high alloy steels

BS spec.	Type	Cu	Si	Mn	Ni	Cr	Mo	C	T N/mm² or M Pa	Yield stress M Pa	Elong. %
3100 BW 10	Austenitic manganese steel	—	1.0	11.0	—	—	—	1.0			
	Possess great hardness and hence is used for earth moving equipment, pinions, sprockets etc. where wear resistance is important.										
3100 410 C21	13% chromium steel	—	1.0	1.0	1.0	13.5	—	0.15	540	370	15
	Mildly corrosion resistant. Used in the paper industry.										
3100 302 C25	Austenitic chromium-nickel steel	—	1.5	2.0	8.0	21.0	—	0.08	480	210	26
	Cast stainless steel. Corrosion resistant and very ductile.										
3100 315 C16	Austenitic chromium-nickel-molybdenum steel	—	1.5	2.0	10.0	20.0	1.0	0.08	480	210	22
	Cast stainless steel with higher nickel content giving increased corrosion resistance. Molybdenum increases weldability.										
3100 302 C35	Heat resisting alloy steel	—	2.0	2.0	10.0	22.0	1.5	0.4	560		3
3100 334 C11		—	3.0	2.0	65.0	10.0	1.0	0.75	460		3
	Can withstand temperatures in excess of 650°C. Temperature at which scaling occurs is raised by increasing amount of chromium.										

Yield stress in N/mm² or MPa T = tensile strength

Cu = copper Si = silicon Mn = manganese Ni = nickel Cr = chromium Mo = molybdenum
C = carbon

Properties of medium and low alloy steels

Type	C	Si	Mn	Cr	Ni	Mo	W	V	T N/mm² or M Pa	Elong. %	Applications etc.
Low alloy structural steel	0.3	0.3	0.75	—	3	—	—	—	800	26	Crankshafts, high tensile shafts etc.
Nickel-chromium-molybdenum steel	0.35	0.3	0.7	0.8	2.8	0.7	—	—	1000	16	Air hardening steel. Used at high temperatures.
High tensile steel	0.4	—	—	1.2	1.5	0.3	—	—	1800	14	Used where high strength is needed.
Spring steel	0.5	1.6	1.3	—	—	—	—	—	1500		
Steel for cutting tools	1.2	—	—	1.5	—	—	4	0.3			
Die steel	0.35	—	0.3	5.0	—	1.4	—	0.4			

W = tungsten   V = vanadium T = tensile strength

Properties of carbon steels to BS970

Type	Composition %			Mechanical properties			Applications etc.
	C	Si	Mn	T N/mm² or MPa	Elong. %	Hardness BHN	
070 M20 (En 2)	0.2	—	0.7	400	21	150	Easily machineable steels suitable for light stressing. Weldable.
070 M26 (En 3)	0.26	—	0.7	430	20	165	Stronger than En 2. Good machineability and is weldable.
080 M30 (En 4)	0.3	—	0.8	460	20	165	Increased carbon increases mechanical properties but slightly less machineable.
080 M36 (En 5)	0.36	—	0.8	490	18	180	Tough steel used for forgings, nuts and bolts, levers, spanners etc.
080 M40 (En 6)	0.4	—	0.8	510	16	180	Medium carbon steel which is readily machineable.
080 M46 (En 8)	0.46	—	0.8	540	14	205	Used for motor shafts, axles, brackets and couplings.
080 M50 (En 9)	0.5	—	0.8	570	14	205	Used where strength is more important than toughness, e.g. machine tool parts.
216 M28 (En 14)	0.28	0.25	1.3	540	10	180	Increased manganese content gives enhanced strength and toughness.
080 M15 (En 32)	0.15	0.25	0.8	460	16	—	Case hardening steel. Used where wear is important, e.g. gears, pawls, etc.

T = tensile strength

Properties of stainless steels

BS ref.	Type	Composition %			Mechanical properties			
		C	Cr	Others	0.2% P.S. N/mm² or MPa	T N/mm² or MPa	Elong. %	Hardness Vickers
410 S21 (En 56A)	Martensitic stainless steel	0.12	13	—	420	590	20	170
431 S29 (En 57)		0.15	16	2.5%Ni	740	900	11	270
	Not suitable for welding or cold forming. Possesses moderate machineability. Used for applications where resistance to tempering at high temperature is important e.g. turbine blades.							
430 S15 (En 60)	Ferritic stainless steel	0.06	16	—	370	540	20	165
	More corrosion resistant than the martensitic steels. They are hardenable by heat treatment. Used for press work because of high ductility.							
302 S25 (En 58A)	Austenitic stainless steel	0.08	18	9.0%Ni	210	510	40	170
	Possesses good resistance to corrosion, good weldability, toughness at low temperature and excellent ductility. May be hardened by cold working.							

T = tensile strength

Properties of high tensile steels

BS ref.	Type	Composition %								Mechanical properties		
		C	Si	Mn	Ni	Cr	Mo	Co	Ti	T N/mm² or MPa	0.2% P.S. N/mm² or MPa	Elong. %
817 M40 (En 24)	Direct hardening nickel steel	0.44	0.35	0.7	1.7	1.4	0.35	—	—	1540	1240	8
970(897M39)	Direct hardening chrome-molybdenum steel	0.35	0.35	0.65	—	3.5	0.7	—	—	1540	1240	10
	Maraging steels	—	—	—	18	—	3.0	8.5	0.20	1480	1400	14

C = carbon   Si = silicon   Mn = manganese   Ni = nickel   Cr = chromium   Mo = molybdenum
Co = cobalt   Ti = titanium T = tensile strength

These steels are used where weight saving is important for instance in the aircraft industry. The deep hardening types are used for plastic moulding dies, shear blades, cold drawing mandrels and pressure vessels. These steels are all difficult to machine.

Aluminium casting alloys

Type	Composition %		Condition	0.2% P.S. N/mm² or M Pa	T N/mm² or M Pa	Elong. %	Hardness BHN	Machine-ability
As cast	Copper	0.1	Sand cast	60	160	5	50	Difficult
	Magnesium	3-6	Chill cast	70	190	7	55	Difficult
	Silicon	10-13						
	Iron	0.6	Die cast	120	280	2	55	—
	Manganese	0.5						
	Nickel	0.1						
	Tin	0.05						
	Lead	0.1						
	Aluminium	balance						
Heat-treatable	Copper	0.7-2.5	Chill cast	100	180	1.5	85	Fair
	Magnesium	0.3						
	Silicon	9.0-11.5	Die cast	150	320	1	85	
	Iron	1.0						
	Manganese	0.5						
	Nickel	1.0						
	Zinc	1.2						
	Aluminium	balance						
	Copper	4-5	Chill cast Fully heat treated	—	300	9	—	Good
	Magnesium	0.1						
	Silicon	0.25						
	Iron	0.25						
	Manganese	0.1						
	Nickel	0.1						
	Zinc	0.1						
	Aluminium	balance						

T = tensile strength

The above alloys are used for food and chemical plant, motor car fittings, marine castings and hydraulic systems.

Properties of wrought aluminium alloys

Type	Composition %	Condition	0.1% P.S. N/mm² or M Pa	T N/mm² or M Pa	Elong. %	Machine-ability	Cold forming
Non-heat treatable alloys	Aluminium 99.99%	Annealed ½ Hard Full hard	— — —	90 max 100–120 130	30 8 5	Poor	Very good
	Copper 0.15 Silicon 0.6 Iron 0.7 Manganese 1.0 Zinc 0.1 Titanium 0.2 Aluminium 97.2	Annealed ¼ Hard ½ Hard ¾ Hard Full hard	— — — — —	115 max 115–145 140–170 160–190 180	30 12 7 5 3	Fair	Very good
	Copper 0.1 Magnesium 7.0 Silicon 0.6 Iron 0.7 Manganese 0.5 Zinc 0.1 Chromium 0.5 Titanium 0.2 Aluminium 90.3	Annealed	90	310–360	18	Good	Fair
Heat treatable alloys	Copper 3.5-4.8 Magnesium 0.6 Silicon 1.5 Iron 1.0 Manganese 1.2 Titanium 0.3 Aluminium balance	Solution treated Fully heat treated	— —	380 420	— —	Good Very good	Good Poor
	Copper 0.1 Magnesium 0.4-1.5 Silicon 0.6-1.3 Iron 0.6 Manganese 0.6 Zinc 0.1 Chromium 0.5 Titanium 0.2 Aluminium balance	Solution treated Fully heat treated	110 230	185 280	18 10	Good Very good	Good Fair

T = tensile strength

Properties of some thermoplastics

Name	Density g/cm³	Tensile strength N/mm² or M Pa	Percentage Elongation at break	E N/mm² or M Pa	Brinell Hardness	Machine-ability
PVC (rigid)	1.33	48	200	3.4	20	Excellent
Polystyrene	1.30	48	3	3.4	25	Fair
PTFE	2.10	13	100	0.3	—	Excellent
Polypropylene	1.20	27	200–700	1.3	10	Excellent
Nylon	1.16	60	90	2.4	10	Excellent
Cellulose nitrate	1.35	48	40	1.4	10	Excellent
Cellulose acetate	1.30	40	10–60	1.4	12	Excellent
Polythene (high density)	1.45	20–30	20–100	0.7	2	Excellent

Properties of some thermosetting plastics

Name	Density g/cm³	Tensile strength N/mm² or M Pa	Percentage Elongation at break	E N/mm² or M Pa	Hardness	Machine-ability
Epoxy resin (glass filled)	1.6–2.0	68–200	4	20	38	Good
Melamine formaldehyde (fabric filled)	1.8–2.0	60–90	—	7	38	Fair
Urea formaldehyde (cellulose filled)	1.5	38–90	1	7–10	51	Fair
Phenol formaldehyde (mica filled)	1.6–1.9	38–50	0.5	17–35	36	Good
Acetals (glass filled)	1.6	58–75	2–7	7	27	Good

Recommendations for the turning of various plastics

Material	Condition	Depth of cut (mm)	Feed (mm/rev)	Cutting speed (m/min)		
				HSS	Brazed carbide	Throw-away carbide tip
Thermoplastics (Polyethylene polypropylene, TFE-fluorcarbon)	Extruded, moulded or cast	4	0.25	50	145	160
High impact styrene and modified acrylic	Extruded moulded, or cast	4	0.25	53	160	175
Nylon, acetals and polycarbonate	Moulded	4	0.25	50	160	175
Polystyrene	Moulded or extruded	4	0.25	18	50	65
Soft grades of thermo-setting plastic	Cast, moulded or filled	4	0.25	50	160	175
Hard grades of thermo-setting plastic	Cast, moulded or filled	4	0.25	48	145	160

Recommendations for the drilling of various plastics

Material	Condition	Cutting speed (m/min)	Feed (mm/rev) Nominal hole diameter (mm)							
			1.5	3.0	6.0	12.0	20.0	25.0	30.0	50.0
Polyethylene, Polypropylene, TFE-fluorocarbon	Extruded, moulded or cast	33	0.12	0.25	0.30	0.38	0.46	0.50	0.64	0.76
High impact styrene and modified acrylic	Extruded, moulded or cast	33	0.05	0.10	0.12	0.15	0.15	0.20	0.20	0.25
Nylon, acetals and Polycarbonate	Moulded	33	0.05	0.12	0.15	0.20	0.25	0.30	0.38	0.38
Polystyrene	Moulded or extruded	66	0.03	0.05	0.08	0.10	0.13	0.15	0.18	0.20
Soft grades of thermo-setting plastic	Cast, moulded or filled	50	0.08	0.13	0.15	0.20	0.25	0.30	0.38	0.38
Hard grades of thermo-setting plastic	Cast, moulded or filled	33	0.05	0.13	0.15	0.20	0.25	0.30	0.38	0.38

Comparison of hardness numbers

Rockwell C Scale	Vickers Pyramid	Brinell	Rockwell C Scale	Vickers Pyramid	Brinell	Rockwell C Scale	Vickers Pyramid	Brinell
68	1030		49	515	468	30	299	286
67	975		48	500	458	29	291	279
66	935		47	485	447	28	284	272
65	895		46	470	436	27	277	266
64	860		45	456	426	26	271	260
63	830		44	442	416	25	265	255
62	800		43	430	406	24	260	250
61	770		42	418	396	23	255	245
60	740		41	406	386	22	250	240
59	715	609	40	395	376	21	245	235
58	690	594	39	385	366	20	240	230
57	670	579	38	375	356		220	210
56	650	564	37	365	346		200	190
55	630	549	36	355	337		180	171
54	610	534	35	345	328		160	152
53	590	519	34	335	319		140	133
52	570	504	33	325	310		120	114
51	550	492	32	315	302		100	95
50	532	480	31	307	294			

Coefficient of thermal conductivity

Substance	Coefft. W/mK	Substance	Coefft. W/mK	Substance	Coefft. W/mK
Silver	420	Mild steel	55	Asbestos	0.18
Copper	386	Cast iron	49	Cork	0.04
Aluminium	202	Lead	36	Felt	0.04
Duralumin	160	Glass	0.8–1.1	Wood	0.02
Brass	120	Concrete	0.9–1.4	Air	0.025
Tin	61	Brick	0.35–0.7		

$$Q = \frac{kAT}{x}$$

Q = rate of heat energy transfer
A = conducting area
T = temperature difference between two faces
x = thickness of material

Calorific values of fuels

Fuel	Calorific Value MJ/kg	Fuel	Calorific Value MJ/kg	Fuel	Calorific Value MJ/kg
Hydrogen	143	Coke	30	Paraffin	46.6
Peat	15	Dry wood	13	Natural gas	52
Carbon	33	Diesel oil	44	Petrol	49
Coal (power station)	22.5–26.5	Refinery residuals	43		
Coal (anthracite)	33–37	Other fuel oils	45–46		

Specific heat capacity of gases

Gas	Constant pressure (c_p)	Constant volume (c_v)	Gas	Constant pressure (c_p)	Constant volume (c_v)
Air	1.005	0.712	Hydrogen	14.486	10.383
Carbon dioxide	0.837	0.653	Nitrogen	1.046	0.754
Carbon monoxide	1.046	0.750	Oxygen	0.921	0.670

Thermo-couples

Type	Low limit °C	High limit °C
Copper/constantan	−250	400
Chromel/constantan	−200	700
Iron/constantan	−200	850
Chromel/alumel	0	1100
Platinum/platinum-rhodium	0	1400
Tungsten/molybdenum	1250	2600

Electrical conductivity

The table below gives the current transmitted by rods 0.10 inches in diameter and 12 inches long when a potential difference of 1 volt is applied across their ends.

Conductors		Conductors		Insulators		Insulators	
Material	Current A	Material	Current A	Material	Current A	Material	Current A
Silver	1080	Iron	166	Wood	10^{-19}	Hard	
Copper	982	Lead	82	Glass	10^{-19}	rubber	10^{-23}
Aluminium	570	Carbon	0.48	Bakelite	10^{-21}	Mica	10^{-20}

Electrolytic corrosion

Material	Potential (volts)	Material	Potential (volts)
Magnesium and its alloys	-1.60	Duralumin	-0.60
Zinc plating on steel	-1.10	Steel	-0.75
Cadmium plating on steel	-0.80	Stainless steel	-0.35
Wrought aluminium alloys	-0.75	Copper and brass	-0.25

The recommendations are:

For parts liable to wetting by sea water or normally exposed to the weather difference of potential should not exceed 0.25 V.

For interior parts which may be exposed to condensation but not to contamination by salt the difference of potential should not exceed 0.50 V.

Dielectric constant

Material	Relative permittivity	Material	Relative permittivity
Air	1.0006	Rubber	2 to 3.5
Paper (dry)	2 to 2.5	Mica	3 to 7
Bakelite	4.5 to 5.5	Porcelain	6 to 7
Glass	5 to 10		

Permittivity of free space $= 8.85 \times 10^{-12}$ F/m

Dielectric strength

Material	Thickness (mm)	Dielectric strength (kV/mm)	Material	Thickness (mm)	Dielectric strength (kV/mm)
Air	1	4.36	Glass (density 2.5)	5	18.3
	6	3.27	Ebonite	1	50.0
	10	2.98	Mica	0.1	61.0
Glass (density 2.5)	1	28.5	Paraffin waxed paper	0.1	50.0

Strength of Materials

$$\sigma = \frac{F}{A}$$

$$\epsilon = \frac{x}{l}$$

$$E = \frac{\sigma}{\epsilon} = \frac{Fl}{Ax}$$

F = applied force or load (N)
A = cross-sectional area (m²)
σ = direct stress (N/m² or Pa)
x = alteration in length
l = original length
ϵ = direct strain (no units)
E = Young's modulus of elasticity (N/m² or Pa)

Percentage elongation $= \dfrac{\text{increase in gauge length}}{\text{original gauge length}} \times 100$

Percentage reduction in area $= \dfrac{\text{original cross-sectional area} - \text{cross-sectional area at fracture}}{\text{original cross-sectional area}} \times 100$

Factor of safety $= \dfrac{\text{ultimate stress}}{\text{working stress}}$

$$G = \frac{\tau}{\gamma}$$

τ = shear stress (N/m² or Pa)
γ = shear strain
G = Modulus of rigidity (N/m² or Pa)

Compound bar:

$$\sigma_1 = \frac{FE_1}{E_1 A_1 + E_2 A_2}$$

$$\sigma_2 = \frac{FE_2}{E_1 A_1 + E_2 A_2}$$

Temperature stress:

$$\sigma = Ea\theta$$

α = coefficient of linear expansion
θ = change in temperature

for a compound bar:

$$(a_1 - a_2)\theta = F\left(\frac{1}{A_1 E_1} + \frac{1}{A_2 E_2}\right)$$

Thin cylindrical shell:

$$\sigma_H = \frac{pd}{2t}$$

$$\sigma_L = \frac{pd}{4t}$$

p = internal pressure (N/m² or Pa)
d = diameter of shell (m)
t = thickness of shell (m)
σ_H = hoop stress (N/m² or Pa)
σ_L = longitudinal stress (N/m² or Pa)

Rotating body:

$$\sigma_H = \rho g \omega^2 r^2 = \rho g v^2$$

ρ = density of material (kg/m³)
ω = angular speed (rad/s)
r = radius (m)
v = linear speed (m/s)

Direct strain energy:

$$U = \frac{\sigma^2}{2E} \times Al$$

Shear stress:

$$\tau = \frac{Q}{A}$$

Q = shearing force (N)

Bearing:

$$F = \sigma_b dt$$

F = force required to cause failure (N)
d = rivet or bolt diameter (m)
t = plate thickness (m)
σ_b = ultimate bearing stress (N/m² or Pa)

Beams Standard Cases

	Maximum B.M.	Maximum Deflection	B.M. Diagram	S.F. Diagram
(cantilever, point load W at end, length l)	Wl	$\dfrac{Wl^3}{3\,EI}$		
(cantilever, UDL total value W, length l)	$\dfrac{Wl}{2}$	$\dfrac{Wl^3}{8\,EI}$		
(simply supported, point load W at centre, length l)	$\dfrac{Wl}{4}$	$\dfrac{Wl^3}{48\,EI}$		
(simply supported, UDL Total value W, length l)	$\dfrac{Wl}{8}$	$\dfrac{5\,Wl^3}{384\,EI}$		

Bending:

$$\frac{M}{I} = \frac{\sigma}{y} = \frac{E}{R}$$

$$Z = \frac{I}{y}$$

$$\sigma = \frac{M}{Z}$$

M = bending moment (Nm)
I = second moment of area (m⁴)
σ = stress due to bending (N/m² or Pa)
y = distance from neutral axis to extreme fibre (m)
Z = section modulus (m³)
R = radius of curvature (m)
E = Young's modulus of elasticity (N/m² or Pa)

Torsion:

$$\frac{\tau}{r} = \frac{G\theta}{l} = \frac{T}{J}$$

θ = angle of twist in radians over a length of l metres
G = modulus of rigidity (N/m² or Pa)
r = radius of shaft (m)
τ = shear stress at radius r (N/m² or Pa)
T = torque (Nm)
J = polar second moment of area (m⁴)

Hardness testing:

$$BHN = \frac{F}{A}$$

$$= \frac{F}{\frac{1}{2}\pi D[D - \sqrt{D - d^2}]}$$

$$VPN = \frac{F}{l^2/2 \sin 68°}$$

$$= \frac{1.854\,F}{l^2}$$

F = applied force (kg)
A = area of indentation (mm²)
D = ball diameter (mm)
d = diameter of indentation (mm)
l = average length of diagonals of indentation (mm)

74

Friction:

$$\mu = \frac{F}{N}$$

$$\tan \phi = \mu$$

μ = coefficient of friction (no units)
F = force tending to cause sliding (N)
N = normal force (N)
ϕ = angle of repose

Coefficients of friction

Surfaces	Coefft.	Surfaces	Coefft.
Metal on metal	0.2	Hardwood on metal—dry	0.6
Rubber on metal	0.4	Hardwood on metal—slightly lubricated	0.2
Leather on metal	0.4	Rubber on a road surface	0.9

The roughness and cleanliness of the surfaces involved has a considerable effect on the coefficient of friction and values much different from the above can easily be obtained.

Friction on an incline plane, force parallel to plane:

(i) When motion is about to occur up the plane
$$P = mg \,(\mu \cos \theta + \sin \theta)$$
(ii) When motion is about to occur down the plane
$$P = (\mu \, mg \cos \theta - \sin \theta)$$

Friction on an incline plane, force horizontal:

(i) When motion is about to occur up the plane
$$P = mg \tan (\theta + \phi)$$
(ii) When motion is about to occur down the plane
$$P = mg \tan (\theta - \phi)$$

θ = angle of inclination of the plane to the horizontal
P = force tending to cause motion (N)

Fluids

$$p = \frac{F}{A}$$

p = pressure (Nm^2)
F = force (N)
A = area (m^2)

$$p = p_a + \rho g h$$

p_a = atmospheric pressure (N/m^2 or Pa)
ρ = density of fluid (kg/m^3)
h = depth of fluid (m)

centre of pressure $= \dfrac{\text{second moment of area about datum}}{\text{first moment of area about datum}}$

$$Q = Av$$

$$A_1 v_1 = A_2 v_2$$

Q = quantity flowing (m^3/s)
A = area of pipe (m^2)
v = speed of flow (m/s)
p = pressure (N/m^2 or Pa)
Z = height from datum (m)

Bernouli's theorem:

$$\frac{p_1}{\rho} + \frac{v_1^2}{2} + gZ_1 = \frac{p_2}{\rho} + \frac{v_2^2}{2} + gZ_2$$

Mechanics

Linear motion:

$$v_f = v_i + at$$
$$s = v_i t + \tfrac{1}{2}at^2$$
$$s = \left(\frac{v_i + v_f}{2}\right)t$$
$$v_f^2 = v_i^2 + 2as$$

v_f = final speed (m/s)
v_i = initial speed (m/s)
a = acceleration (m/s²)
t = time (s)
s = distance (m)

Circular motion:

$$\omega_f = \omega_i + at$$
$$\theta = \omega_i t + \tfrac{1}{2}at^2$$
$$\theta = \left(\frac{\omega_i + \omega_f}{2}\right)t$$
$$\omega_f^2 = \omega_i^2 + 2a\theta$$

ω_f = final speed (rad/s)
ω_i = initial speed (rad/s)
a = angular acceleration (rad/s²)
t = time(s)
θ = angle turned (rad)

Centripetal acceleration $= \dfrac{v^2}{r} = \omega^2 r$

Centrifugal force $= \dfrac{mv^2}{r} = m\omega^2 r$

m = mass rotating (kg)
v = tangential speed (m/s)
r = radius (m)
ω = circular speed (rad/s)

Momentum $= mv$

$$F = ma$$
$$W = mg = 9.81 \text{ m}$$

F = force causing acceleration (N)
W = weight of body (N)

Force of a jet of water on a stationary flat plate:

$$F = \rho A v^2$$

ρ = density (kg/m³)
A = area of jet (m²)
v = speed of jet (m/s)

Force of a jet of water on a semi-circular stationary vane:

$$F = 2\rho A v^2$$

Work done:

$$W = Fs$$

$$W = T\theta$$

W = work done (J)
F = force moving object (N)
s = distance moved in direction of force (m)
T = torque (Nm)
θ = angle turned (rad)

Power:

$$P = \frac{Fs}{t} = Fv = \frac{T\theta}{t} = 2\pi NT$$

P = power (W)
v = velocity (m/s)
N = circular speed (rev/s)

Potential energy $(W) = mgh$

Kinetic energy $(W) = \tfrac{1}{2}mv^2$

Change in kinetic energy $(W) = \tfrac{1}{2}m(v_f^2 - v_i^2)$

g = acceleration due to gravity (m²/s)
m = mass (kg)
h = height of body above datum (m)

Falling bodies:

$$v = \sqrt{2gh}$$

v = final speed (m/s)
h = height fallen through (m)

Impact:

$$m_A v_A + m_B v_B = m_A v'_A + m_B v'_B$$
$$e = \frac{v'_A - v'_B}{v_A - v_B}$$

v = velocity before impact
v' = velocity after impact
e = coefficient of restitution

Linear impulse:

$$Ft = m(v_f - v_i)$$

Vehicle on a curved horizontal track:

$$v = \sqrt{g\mu r} \text{ (skidding)}$$

$$v = \sqrt{\frac{gar}{h}} \text{ (overturning)}$$

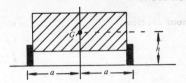

Vehicle on a curved banked track:

$$v = \sqrt{gr\left(\frac{\mu + \tan\theta}{1 - \mu\tan\theta}\right)}$$

v = limiting speed for skidding

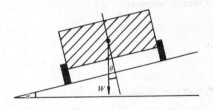

Torque:

$$T = I\alpha$$
$$I = mk^2$$

T = torque (Nm)
I = moment of inertia (kg/m²)
α = angular acceleration (rad/s²)
ω = angular speed (rad/s)
k = radius of gyration (m)
m = mass (kg)

Angular momentum $= I\omega$

Kinetic energy of rotation $= \frac{1}{2}I\omega^2$

Simple harmonic motion (SHM):

$$T = \frac{2\pi}{\omega}$$

$$f = \frac{1}{T} = \frac{\omega}{2\pi}$$

$$\frac{\ddot{x}}{x} = -\omega^2$$

T = periodic time (s)
f = frequency (Hz)
$\ddot{x}$ = acceleration
x = displacement
ω = circular speed (rad/s)

or
$$\text{Acceleration} = -\omega^2 \times \text{displacement}$$

for a spring:

$$f = \frac{1}{2\pi}\sqrt{\frac{k}{m}}$$

$$f = \frac{1}{2\pi}\sqrt{\frac{g}{\delta_s}}$$

$$\delta_s = \frac{mg}{k}$$

k = spring stiffness (N/m)
m = mass on spring (kg)
δ_s = deflection of spring under mass (m)

Simple pendulum:

$$f = \frac{1}{2\pi}\sqrt{\frac{g}{l}}$$

l = length of pendulum (m)

Compound pendulum:

$$T = 2\pi\sqrt{\frac{a^2 + k_G^2}{ag}}$$

k_G = radius of gyration through centre of mass (m)
a = distance from point of suspension to G (m)

Conical pendulum:

$$T = ml\omega^2$$

T = tension in string (N)
m = mass of bob (kg)
ω = angular speed (rad/s)

Machines

Mechanical advantage or force ratio $= \dfrac{\text{load}}{\text{effort}}$

Velocity ratio or movement ratio $= \dfrac{\text{distance moved by effort}}{\text{distance moved by load}}$

Efficiency $= \eta = \dfrac{\text{useful work out}}{\text{work put in}} = \dfrac{\text{force ratio}}{\text{movement ratio}}$

Movement ratio of wheel and axle $= \dfrac{\text{radius of wheel}}{\text{radius of axle}}$

Movement ratio of wheel and differential axle

$$= \dfrac{2 \times \text{radius of wheel}}{\text{radius of larger part of axle} - \text{radius of smaller part}}$$

Movement ratio of screw jack $= \dfrac{2\pi \times \text{radius at which effort is applied}}{\text{lead of screw}}$

Movement ratio of gear drive $= \dfrac{\text{number of teeth on driven wheel}}{\text{number of teeth on driver wheel}}$

For a gear drive: $\dfrac{N_A}{N_B} = \dfrac{n_B}{n_A}$ 　　　　N = rotation speed (rev/min)
　　　　　　　　　　　　　　　　　　n = number of teeth on gear

For a belt drive: $\dfrac{N_A}{N_B} = \dfrac{d_B}{d_A}$ 　　　　d = pulley diameter

Power of a belt drive:

$$P = v(F_1 - F_2)$$

$$F_1 = F_2\, e^{\mu\theta} \text{ for flat belt}$$

$$F_1 = F_2\, e^{\mu\theta/\text{cosec}\,\alpha} \text{ for V-belt}$$

F_1 = tension on tight side (N)
F_2 = tension on slack side (N)
v = linear speed of belt (m/s)
P = power developed (W)
θ = angle of lap (radians)
μ = coefficient of friction
α = angle of vee

Spirit level: θ (radians) $= \dfrac{\text{movement of bubble}}{R}$

θ (degrees) $= \dfrac{57.3 \times \text{movement of bubble}}{R}$

Sine bar: $h = C \sin \alpha$

Measurement of large bores: $D = L + \dfrac{\omega^2}{2L}$

Measurement of screw threads: $W = D_E + d(1 + \text{cosec}\,\tfrac{\alpha}{2}) - \tfrac{p}{2}\cot\tfrac{\alpha}{2}$

for metric threads: $W = D + 3d - 1.516\,p$

best wire size is: $d = 0.577\,p$

Tool life: $ST^n = c$

(for H.S.S. tools $n = \tfrac{1}{7}$ to $\tfrac{1}{8}$ for roughing cuts in steel and $\tfrac{1}{10}$ for light cuts in steel. $n = \tfrac{1}{5}$ when using a tungsten carbide tool for roughing cuts on steel)

R = vial radius
θ = angle of inclination of spirit level
h = difference in height of plugs
C = centre distance of plugs
α = angular setting of sine bar
D_E = bore diameter
L = length of gauge
ω = half total amount of rock
D_2 = effective diameter
d = wire diameter
α = flank angle
p = pitch
W = diameter over wires
D = major diameter
S = cutting speed (m/min)
T = tool life (min)
n = constant
c = constant

Change wheels: $\dfrac{\text{number of teeth in driving gears}}{\text{number of teeth in driven wheels}} = \dfrac{\text{lead of thread to be cut}}{\text{pitch of leadscrew}}$

Dividing head:

Simple indexing $= \dfrac{40}{n}$ $\qquad$ n = required number of divisions

Angular indexing $= \dfrac{\theta}{9}$ $\qquad$ θ = angle required

Cutting speeds in metres per minute

Material	Tool material			
	H.S.S.	Super H.S.S.	Stellite	Tungsten Carbide
Aluminium alloys	70–100	90–120	over 200	over 350
Brass (free cutting)	70–100	90–120	170–250	350–500
Bronze	40–70	50–80	70–150	150–250
Grey cast iron	35–50	45–60	60–90	90–120
Copper	35–70	50–90	70–150	100–300
Magnesium alloys	85–135	110–150	85–135	85–135
Monel metal	15–20	18–25	25–45	50–80
Mild steel	35–50	45–60	70–120	—
High tensile steel	5–10	7–12	20–35	—
Stainless steel	10–15	12–18	30–50	—
Thermo-setting plastic	35–50	45–60	70–120	100–200

$$S = \frac{\pi d N}{1000}$$

S = cutting speed (m/min)
d = work or cutter diameter (mm)
N = rotational speed (rev/min)

Power used in cutting

Material	k_L N/mm²	k_d	k_m J/mm³	Material	k_L N/mm²	k_d	k_m J/mm³
Aluminium	700	0.11	0.9	Mild steel	1200	0.36	2.7
Brass	1250	0.084	1.6	Tool steel	3000	0.4	7.0
Cast iron	900	0.07	1.9				

For lathework: $\qquad P = \dfrac{k_L d f S}{60\,000}$

For drilling: $\qquad T = k_d f^{0.75} D^{1.8}$

$\qquad\qquad\qquad P = \dfrac{2\pi N T}{60\,000}$

For milling: $\qquad P = \dfrac{k_m d w f_m}{60}$

P = power used (kw)
d = depth of cut (mm)
f = feed (mm/rev)
S = cutting speed (m/min)
T = torque (Nm)
D = drill diameter (mm)
N = rotational speed (rev/min)
W = width of cut (mm)
f_m = milling machine table feed (mm/min)
V = Volume of metal removed (cm³/min)

Volume of metal removed:

Lathework: $\qquad V = dfS$

Drilling: $\qquad V = \dfrac{\pi D^2}{4} f N$

Milling: $\qquad V = \dfrac{W d f_m}{1000}$

Electrical Laws, Definitions and Formulae

The ampere: That current which, if maintained in two straight parallel conductors of infinite length, of negligible circular cross-section, and placed 1 metre apart in a vacuum, would produce between these conductors a force of 2×10^{-7} newtons per metre length. Symbol I.

Ohms' Law: The current in a circuit is directly proportional to the voltage and inversely proportional to the resistance providing the temperature remains constant.

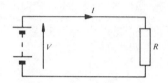

$$I = \frac{V}{R} \text{ amperes}$$

$$V = IR \text{ volts}$$

$$R = \frac{V}{I} \text{ ohms}$$

$P = VI$ watts $\qquad$ $W = VIt$ joules

$P = I^2R$ watts $\qquad\quad$ $= I^2Rt$ joules

$P = \dfrac{V^2}{R}$ watts $\qquad$ $= \dfrac{V^2t}{R}$ joules

$Q = It$ coulombs

$1 \text{ kWh} = 1 \text{ Board of Trade Unit (unit)} = 3.6 \times 10^6 \text{ joules}$

Resistivity: The resistance measured across the two opposite faces of a unit cube of material at a particular temperature. Symbol ρ (rho).

$R = \dfrac{\rho l}{a}$ where ρ = Resistivity, l = Length, a = cross-sectional area.

Resistors in series

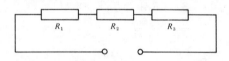

Total resistance $R_\text{T} = R_1 + R_2 + R_3$

Resistors in parallel

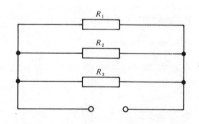

$$\frac{1}{R_\text{T}} = \frac{1}{R_1} + \frac{1}{R_2} + \frac{1}{R_3}$$

Potential divider

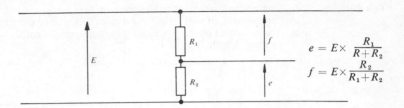

$$e = E \times \frac{R_1}{R+R_2}$$

$$f = E \times \frac{R_2}{R_1+R_2}$$

Kirchoff's Law No. 1

The algebraic sum of the currents meeting at a point is zero.

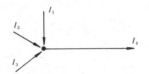

$$I_4 = I_1+I_2+I_3$$

Kirchoff's Law No. 2

The sum of the internal potential differences in a circuit equals the applied e.m.f.

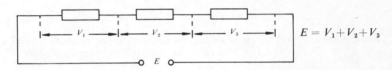

$$E = V_1+V_2+V_3$$

Temperature coefficient of resistance

The ratio of the increase (or decrease) of resistance per degree rise of temperature to the resistance at 0°C. Symbol a (alpha).

1. $R_T = R_0(1+a\theta)$ General formula

2. $\dfrac{R_1}{R_2} = \dfrac{1+a\theta_1}{1+a\theta_2}$ θ = temperature in °C

Lenz's Law

The direction of the induced e.m.f. always produces a current in a direction such as to oppose the motion or change of flux responsible for inducing that e.m.f.

The henry

If in a circuit an e.m.f. of 1V is induced when the current changes at the rate of 1 A/s then the inductance is 1 henry. Symbol L.

Permeability of free space $= \mu_0 = 4\pi \times 10^{-7}$ H/m   Relative permeability $= \mu_r$

Absolute permeability $= \mu = \mu_0 \mu_r = \dfrac{B}{H}$ B = Flux density $= \dfrac{\Phi}{a}$ teslas

H = Magnetic field strength $= \dfrac{IN}{l}$ amperes/metre N = number of turns

l = length of magnetic circuit

Reluctance $= S = \dfrac{l}{\mu_0 \mu_r a}$

Magnetic energy stored in non-magnetic medium $\left.\right\} = \dfrac{B^2}{2\mu_0}$ joules/m^3

Magnetic pull between two iron surfaces $\left.\right\} = \dfrac{B^2 a}{2\mu_0}$ newtons

d.c. generators

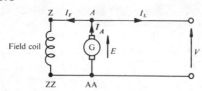

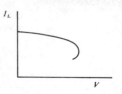

Shunt connected

$$I_A = I_L + I_F$$
$$E = V + I_a R_a$$
$$E = \frac{2P\Phi NZ}{a}$$

R = Armature resistance
P = Pair of poles
Φ = Flux per pole in webers
N = Speed in rev/s
Z = Number of armature conductors
a = $2P$ for lap wound and 2 for wave wound armature

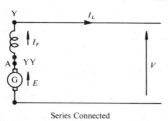

Series Connected

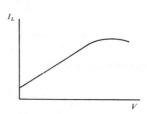

Series Connected

$$I_L = I_a = I_F$$

d.c. motors

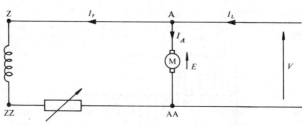

$$I_A = I_L - I_F \qquad V = E_B + I_A R_a$$

$$E_B = \frac{2P\Phi NZ}{a} \text{ (See \textbf{Generator}) } E_B \text{ is the ``Back e.m.f.''}$$

To reverse direction, reverse connections A–AA *or* reverse connection Z–ZZ but not both.

Torque $= 0.318\frac{I_a}{c}ZP\Phi$ newton metres   Torque a ΦI_a

$E = L \times$ rate of change of current

$$= L \times \frac{(I_2 - I_1)}{t}$$

$$= L\frac{di}{dt} \text{ volts}$$

Also $E = N \times$ rate of change of flux

$$= N \times \frac{(\Phi_2 - \Phi_1)}{t}$$

$$= N\frac{d\Phi}{dt}$$

Energy stored $\mathscr{W} = \tfrac{1}{2}LI^2$ joules

The farad (F)

If in a circuit a quantity of electricity of 1 coulomb is stored when an e.m.f. of 1 volt is applied then the capacity is 1 farad.

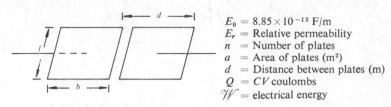

$E_0 = 8.85 \times 10^{-12}$ F/m
E_r = Relative permeability
n = Number of plates
a = Area of plates (m²)
d = Distance between plates (m)
$Q = CV$ coulombs
$\mathcal{W}$ = electrical energy

$$a = l \times b$$
$$C = \frac{E_0 E_r (n-1)a}{d} \text{ farads}$$

Energy stored $\mathcal{W} = \frac{1}{2}CV^2$ joules

Capacitors in series

To find total capacitance

$$\frac{1}{C_T} = \frac{1}{C_1} + \frac{1}{C_2} + \frac{1}{C_3}$$

Capacitors in parallel

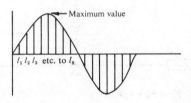

Total capacitance $= C_T = C_1 + C_2 + C_3$

a.c. theory

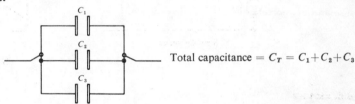

Maximum value

$I_1 I_2 I_3$ etc. to I_8

Average value of an alternating current $= \dfrac{I_1 + I_2 + I_3 + I_4 + \ldots I_n}{n}$

For a sine wave I average $= \dfrac{1}{\pi} \displaystyle\int_0^\pi \sin x \, dx = 0.636 \times$ max. value.

Root mean square (r.m.s.) value $= \sqrt{\dfrac{I_1^2 + I_2^2 + I_3^2 + I_4^2 + \ldots I_n^2}{n}}$

For a sine wave $= \sqrt{\dfrac{1}{2\pi} \displaystyle\int_0^{2\pi} \sin^2 x \, dx} = 0.707 \times$ max. value.

Peak or crest factor $= \dfrac{\text{peak value}}{\text{r.m.s. value}} = 1.414$ for a sine wave.

Form factor $= \dfrac{\text{r.m.s. value}}{\text{average value}} = 1.11$ for a sine wave.

a.c. and resistance

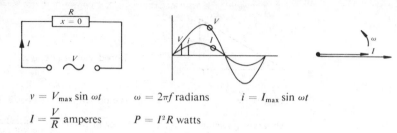

$$v = V_{max} \sin \omega t \qquad \omega = 2\pi f \text{ radians} \qquad i = I_{max} \sin \omega t$$

$$I = \frac{V}{R} \text{ amperes} \qquad P = I^2 R \text{ watts}$$

a.c. and inductance

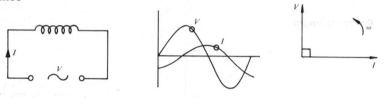

Inductive reactance $X_L = 2\pi f L$ ohms

a.c. and capacitance

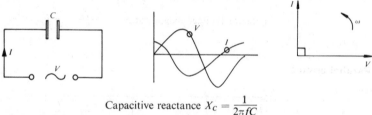

Capacitive reactance $X_C = \dfrac{1}{2\pi f C}$

a.c. general series circuit

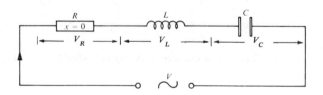

$$V_R = I \times R, \quad V_L = I \times X_L, \quad V_C = I \times X_C, \quad V = \sqrt{V_R^2 + (V_L - V_C)^2}$$

Phasor diagram

As it is a series circuit I is taken as reference

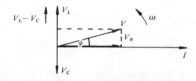

ϕ is the phase angle

Impedance triangle

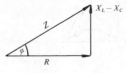

Impedance of circuit $Z = \sqrt{R^2 + (X_L - X_C)^2}$

$\cos \phi = \dfrac{R}{Z}$

True power $= V \times I \times \cos \phi$ watts

Apparent power $=$
$V \times I$ volts amperes

Power factor $= \dfrac{\text{True power}}{\text{Apparent power}} = \cos \phi$

Power triangle

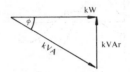

When $V_L = V_c$ and $X_L = X_c$, then $Z = R$
and resonance occurs. Power factor is unity.

Resonant frequency $f_r = \dfrac{1}{2\pi\sqrt{LC}}$ Hz

Q factor (voltage magnification) $= \dfrac{2\pi f_L}{R} = \dfrac{1}{R}\sqrt{\dfrac{L}{C}} = \dfrac{V_L}{V}$

Parallel circuit

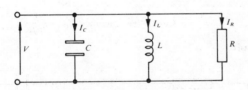

For phasor diagram, V is taken as reference.

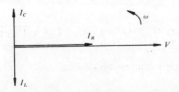

If $I_c = I_L$ then resonance occurs

$f_r = \dfrac{1}{2\pi L}\sqrt{\dfrac{L}{C} - R^2}$ if $R \ll 2\pi f_L$ then $f_r = \dfrac{1}{2\pi}\sqrt{LC}$

Dynamic impedance $= \dfrac{L}{CR}$ Current magnification $Q = \dfrac{2\pi f_L}{R} = \dfrac{I_L}{I}$

Three-phase a.c. circuits
Star

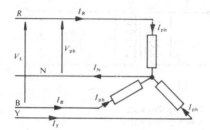

In star $I_{ph} = I_L$ $V_L = \sqrt{3}\, V_{ph}$

Neutral current I_N is the phasor sum of the three line current I_R, I_Y and I_B

$$I_N = I_R + I_Y + I_B$$

For balanced conditions $I_N = 0$

Mesh or delta

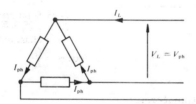

In delta $V_L = V_{ph}$
$I_L = \sqrt{3}\, I_{ph}$

Three-phase power

$$P = \sqrt{3}\, V_L I_L \cos \phi$$

Two-wattmeter method

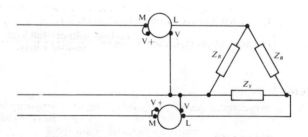

Total power $= W_1 + W_2$ $\tan \phi = \sqrt{3}\left(\dfrac{W_2 - W_1}{W_2 + W_1}\right)$

a.c. machines:
Transformer

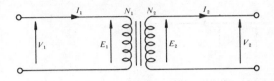

e.m.f. $E_1 = 4.44N_1 f \Phi m$ volts
$E_2 = 4.44N_2 f \Phi m$ volts

Turns ratio $\dfrac{V_1}{V_2} \sim \dfrac{N_1}{N_2} \sim \dfrac{I_2}{I_1}$

Phasor diagram for 1:1 transformer on no-load

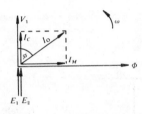

I_M = Primary current component
producing magnetic field

I_c = Primary current component
supplying hysteresis and core
loss

I_0 = Total primary current

$I_0 = \sqrt{I^2{}_c + I^2{}_M}$

ϕ = Phase angle

$\cos \phi = \dfrac{I_c}{I_0}$

Phasor diagram for a 1:1 loaded transformer

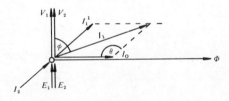

I_2 = Secondary current I_1 = Primary current due to secondary current
I_1 = Total primary current   By cosine rule $I_1{}^2 = I_0{}^2 + I_1 - 2I_0 I_1 \cos \theta$

Per unit voltage regulation $= \dfrac{\text{no-load voltage} - \text{full load voltage}}{\text{no-load voltage}}$

Induction Motor

Synchronous speed $N_s = \dfrac{f}{p}$ rev/s   where f = frequency in hertz
p = pairs of poles

Per unit slip $= \dfrac{\text{synchronous speed} - \text{rotor speed}}{\text{synchronous speed}}$

$S = \dfrac{N_s - N_r}{N_s}$ Rotor frequency $(f_r) = sf$

Electronics
Triodes

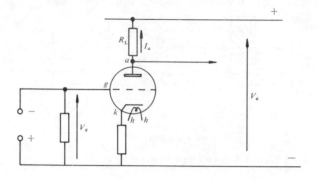

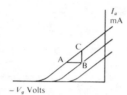

$$AB = \triangle V_g \quad \text{change in } V_g$$
$$BC = \triangle I_a \quad \text{change in } I_a$$

$$\text{Mutual conductance } g_m = \frac{\triangle I_a}{\triangle V_g} \text{ siemens}$$

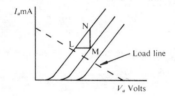

$$LM = \triangle V_a \qquad MN = \triangle I_a$$

$$\text{Anode (a.c.) resistance} = r_a = \frac{\triangle V_a}{\triangle I_a}$$

$$\text{Amplification factor } \mu = g_m \, r_a \qquad r_a = \frac{\triangle V_a}{\triangle V_g} \qquad \text{Gain} = \frac{\mu R_L}{r_a + R_L}$$

Cathode ray tubes

Force on an electron moving across a magnetic field $= Bqv$ newtons where B is the flux density in teslas, q is the negative charge on each electron measured in coulombs and v is the velocity in m/s.

Electro-static deflection

$$D = \frac{Ll}{2d} \times \frac{V}{V_a} \text{ metres}$$

V is the p.d. between deflecting plates
V_a is the voltage on the final
    accelerating anode

Electromagnetic deflection

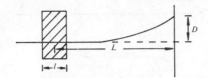

$$D = BLl\sqrt{\frac{q}{m}} \times \frac{1}{2V_a} \text{ metres}$$

q = Negative charge on electron in coulombs
m = Mass of electron in kilogrammes
V_a = Voltage on final accelerating anode

Transistor:
Common base

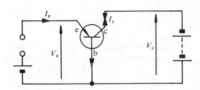

Current amplification factor $\alpha = \dfrac{\triangle I_c}{\triangle I_e}$

Input resistance $R_{\text{in}} = \dfrac{\triangle V_e}{\triangle I_e}\,\Omega$

Output resistance $R_{\text{out}} = \dfrac{\triangle V_c}{\triangle I_c}\,\Omega$

Resistance gain $= \dfrac{R_{\text{out}}}{R_{\text{in}}}$ Voltage gain $A = \alpha \times$ resistance gain

Power gain $P_g = \alpha \times A$

Common emitter

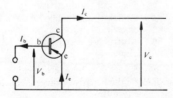

Current amplification factor $(\alpha') = \dfrac{\triangle I_c}{\triangle I_b}$

$\alpha' = \dfrac{\alpha}{1-\alpha}$ $R_{\text{out}} = \dfrac{\triangle V_c}{\triangle I_c}$ $R_{\text{in}} = \dfrac{\triangle V_b}{\triangle I_b}$

Index